AA002361

**MATERIALS RESEARCH SOCIETY
SYMPOSIUM PROCEEDINGS VOLUME 825**

Semiconductor Spintronics

April 12-16, 2004
San Francisco, California, USA

Printed from e-media with permission by:

Curran Associates, Inc.
57 Morehouse Lane
Red Hook, NY 12571
www.proceedings.com

ISBN: 1-55899-753-9

Some format issues inherent in the e-media version may also appear in this print version.

CAMBRIDGE UNIVERSITY PRESS
Cambridge, New York, Melbourne, Madrid, Cape Town,
Singapore, São Paulo, Delhi, Tokyo, Mexico City

Cambridge University Press
32 Avenue of the Americas, New York, NY 10013-2473, USA

www.cambridge.org

Materials Research Society
506 Keystone Drive, Warrendale, PA 15086
http://www.mrs.org

©Materials Research Society 2004

This publication is in copyright. Subject to statutory exception
and to the provisions of relevant collective licensing agreements,
no reproduction of any part may take place without the written
permission of Cambridge University Press.

First published 2004

CODEN: MRSPDH

ISBN: 1-55899-753-9

Cambridge University Press has no responsibility for the persistence or
accuracy of URLs for external or third-part Internet Web sites referred to
in this publication and does not guarantee that any content on such Web sites
is, or will remain, accurate or appropriate.

Additional copies of this publication are available from:

Curran Associates, Inc.
57 Morehouse Lane
Red Hook, NY 12571 USA
Phone: 845-758-0400
Fax: 845-758-2634
Email: curran@proceedings.com
Web: www.proceedings.com

Semiconductor Spintronics

Materials Research Society Symposium Proceedings Volume 825

San Francisco, California, USA
12-16 April 2004

TABLE OF CONTENTS

1 **Investigation of Magnetic Properties in Mn Incorporated InSb, InP, and GaAs, Synthesized Through Controlled-Ambient Annealing**
Hollingsworth, Joel;Bandaru, Prabhakar

7 **Optical Properties of Transition Metal Doped ZnO Ceramics and Thin Films**
Katiyar, Ram S.;Awasthi, Neha;Bhattacharya, Pijush

13 **Valencies of Mn Impurities in ZnO**
Petit, Leon;Schulthess, Thomas C.;Svane, Axel;Temmerman, Walter M.;Szotek, Zdzislawa

19 **Model for Spin Injection Into Conjugated Organic Semiconductors**
Ruden, P. Paul;Smith, Darryl L.

25 **Spin Effects of Low-Dimensional Electron Gases Studied by Far-Infrared Photoconductivity Experiments**
Hu, C.-M.

37 **Spin-Orbit Coupling and Magnetic Spin States in Cylindrical Quantum Dots**
Marques, Gilmar Eugenio

43 **Theory of Electrically Controlled Resonant Tunneling Spin Devices**
Ting, David Z.-Y.;Cartoixa, Xavier

49 **Growth of Mirror-Like Zn1-XMnxO Diluted Magnetic Semiconductor Thin Films by R.f. Magnetron Sputtering Method**
Lee, Sejoon;Lee, Hye Sung;Kim, Deuk Young

55 **Electronic Structure of the Diluted Magnetic Semiconductors Pb1-XSnxTe:Yb**
Skipetrov, Evgenii P.;Zvereva, Elena A.;Volkova, Olga;Golubev, Alexander;Slyn'Ko, Vasilii

61 **Magnetic Properties of GaN Layers Implanted by Mn, Cr or V**
Guzenko, Vitaliy A.;Thillosen, Nicolas;Dahmen, Andre;Calarco, Raffaella;Schapers, Thomas;Luysberg, Martina;Houben, Lothar

67 **Optimization of Sample Holder Materials for Sensitive Magnetometry Measurements at Low Temperatures**
Bossi, I.;Dilley, N. R.;O'Brien, J. R.;Spagna, S.

73 **Rashba Spin-Orbit Coupling in InGaAs/InP Quantum Wires**
Schapers, Thomas;Guzenko, Vitaliy A.;Knobbe, Jens

79 **Selective Voltage-Controlled Hole Spin in Non-Magnetic Resonant Tunneling Diodes**
Marques, Gilmar Eugenio

85 **Spin Lifetime Tuning in Zincblende Heterostructures and Applications to Spin Devices**
Cartoixa, Xavier;Ting, David Z.-Y.;Chang, Yia-Chung

Investigation of magnetic properties in Mn incorporated InSb, InP, and GaAs, synthesized through controlled-ambient annealing

Joel Hollingsworth, Prabhakar Bandaru
Materials Science Program, Mechanical & Aerospace Engineering department,
U.C. San Diego, La Jolla, CA, 92093

ABSTRACT

Magnetic semiconductors are of interest for emerging spintronic applications, such as the integration of electronic information processing with magnetic data storage. We report on a new approach - furnace annealing under controlled ambients – aimed at increasing Mn incorporation and synthesizing new magnetic semiconductors with T_c greater than/around room temperature. These annealing treatments are hypothesized to reduce the effect of Mn interstitials. We have obtained preliminary SQUID magnetometry results which indicate ferromagnetic Curie temperatures of around 130 K in (In,Mn) Sb and 60 K in (In, Mn)P. X-ray diffraction was used to characterize phase homogeneity.

INTRODUCTION

The recent discovery of magnetism in non-magnetic semiconductors such as GaAs[1] and Ge[2] has spurred inquiry into the possibility of harnessing the electron spin in the solid state for the creation of a whole new class of technologies and devices. These include spin transistors[3], the solid-state analogs of the electro-optic modulator, spin-LEDs[4] (Light Emitting Diodes) that can emit polarized light, and spin-based quantum logic[5]. Another most intriguing aspect is the possibility of electrically varying magnetic character, as seen in Mn doped InAs[6].

According to currently accepted models, ferromagnetism arises in magnetic semiconductors through Zener type[7] interactions where the holes introduced through Mn doping mediate the Mn moment alignment. This mechanism is closely related to the RKKY (Ruderman-Kittel-Kasuya-Yosida) interaction[8] where the mutual anti-ferromagnetic alignment between the carrier (hole) and the Mn ion, can cause two separated Mn ions to be ferromagnetically oriented. This approach predicates that ferromagnetism (and the Curie temperature: T_c) in III-V semiconductors where Mn is an acceptor, is proportional to the amount of Mn^{2+} that can be incorporated, and the number of holes introduced.

However, control of the mechanism of Mn incorporation is critical to this endeavor. Prior efforts to increase T_c in (Ga,Mn)As have focused on introducing concentrations of Mn well above the thermodynamic solubility limit (p: $1.4 \cdot 10^{20}/cm^3$)[9], by such means as low temperature MBE[1] and the growth of digital structures, with alternating Mn and GaAs layers[10]. However, Mn has been shown both to substitute for Ga (Mn_{Ga}) and also to reside in the crystal's interstices (Mn_i). As lattice sites are filled, Mn can reasonably be expected to preferentially fill the more abundant interstitial sites. Unfortunately, Mn_i, along with intrinsic point defects, e.g. As_{Ga} anti-site defects, are donor like and compensate[11] substitutional Mn_{Ga}. Additionally, Mn is seen to precipitate out when its solubility limit is reached, either in an elemental form or as Manganese Arsenides (MnAs). As a consequence, low-temperature methods have not succeeded in increasing T_c. What did succeed in raising T_c to 160 K, from an "intrinsic limit" of 110 K[12], was a simple process of annealing and subsequent homogenization[13]; the maximum annealing temperature being restricted to the temperature of MBE growth (~ 525 K). The greater Tc was attributed to the destruction of Mn_i and overall homogenization[14]; however, these processes again seem to result in the removal of Mn.

It has been shown earlier[15, 16], that Mn concentrations of up to $10^{20}/cm^3$, comparable to those currently obtained by MBE, can be achieved using diffusion and annealing processes. Additionally, solid-state reactions at high temperatures (> 1300 K) have been used to prepare Mn and Co doped ZnO[17] and (Ga, Mn) As[18]. Ion implantation and subsequent annealing have produced InP: Mn with T_c of 90 K[19], and InSb: Mn exhibiting T_c of 8.5 K has been produced by MBE[20], suggesting that more work can be done on these materials.

Prior work has often centered on MBE and other capital-intensive methods, but an alternate, more economical synthesis method might allow a faster and more extensive survey of possible compositions. In the present work, we use ampoule annealing as a means of exploring less well-studied magnetic semiconductor compositions. Additionally, this method allows a wide range of processing environments and temperatures, which can be tailored to influence the equilibrium solubility and site occupancy of Mn. For instance, high anion overpressure is expected to increase cation vacancy concentration and thereby promote both Mn solubility and substitutional occupancy (such as Mn_{As}).

Ampoule annealing is also well suited to co-doping experiments; as an example of this, GaAs was annealed with both Mn and Cr. Cr was chosen because of its larger affinity[21] (higher crystal field stabilization energy: CFSE) for occupying the tetrahedral sites compared to Mn^{2+} (CFSE: 0.00 eV), viz. Cr^{3+}(0.69 eV). From ligand field considerations[21], Cr also favors the high co-ordination hexagonal interstitial sites when substituted into GaAs, precluding Mn_i formation. Cr is also expected to partially compensate for the formation of As_{Ga} mid-gap deep-level states[22].

Based on the proposed rationale, furnace annealing, as an alternative to more expensive MBE based synthesis, was used to explore less well-studied magnetic semiconductor compositions and processing conditions, including excess anion (P, Sb and As) pressure and the addition of elemental Cr.

EXPERIMENTAL DETAILS

Semi-insulating GaAs ($\rho \sim 10^8 \, \Omega$ cm), InP ($\rho \sim 10^7 \, \Omega$ cm), and InSb ($\rho \sim 10^5 \, \Omega$ cm) (100) wafers, 600 microns thick, (University wafers Inc.) were used for the experiments. These non-magnetic semiconductors were ultrasonically cleaned in organic and inorganic reagents and placed along with Mn and Cr (99.99% purity) in sealed quartz ampoules in an electric furnace. The annealing was performed at various temperatures in the range of 600 - 1000 K and for times, 1-100 hours. Additionally, different weights of elemental P or As (99.999% purity) were placed in the InP and GaAs ampoules respectively to vary the ambient pressure between 0.1-10 atmospheres.

X-ray diffractometry (Philips X-Pert, Rigaku) and $\theta - 2\theta$ scans, using Cu Kα radiation (λ: 1.542 Å), at room temperature was used to determine the lattice parameters and characterize the crystal quality, the presence of impurity phases, and strain in the Mn incorporated semiconductors. The shift in the lattice parameter was used to estimate the amount of Mn substitution, assuming a Mn-Sb bond length of 2.69 Å[23], and a Vegard's law dependence of the lattice parameter on Mn concentration. At $2\theta = 24^\circ$, the diffractometry signal should come from a layer 13μm thick. From the published values of the diffusion constant (D) for Mn into GaAs[23] at similar homologous temperature, we estimate D for Mn penetration into InSb to be of the order of $10^{-10}\,cm^2$/sec. Consequently, for our annealing conditions, Mn would penetrate around 60 μm, which is several times the x-ray penetration depth; this suggests that the unmodified substrate will not produce an appreciable diffraction signal near this angle.

A Superconducting Quantum Interference Device (SQUID, Quantum Design, Inc.) magnetometer was used to measure the Magnetization (M) – Temperature (T) characteristics and estimate T_c. The samples were cooled from room temperature to 5 K, and a magnetic bias field of 0.1 Tesla was then applied in the plane of the film. Hysteresis loops (M-H) indicated the presence of ferromagnetic/ferrimagnetic phases.

EXPERIMENTAL RESULTS AND DISCUSSION

Magnetic Characterization: SQUID Magnetometry

The observed magnetization, in emu (electromagnetic units), has been normalized by the approximate surface area of the sample. We present preliminary results representative of the wide variety of magnetic behavior seen in Mn doped InSb, InP and GaAs.

InSb, when annealed along with Mn, indicates a Weiss mean field-like behavior for the M vs. T characteristic (Figure 1a), which is indicative of a metallic nature for this particular dilute magnetic semiconductor[25] system. An approximate T_c around 130 K is obtained through extrapolation, which is an order of magnitude higher compared to prior results[20]. The magnetization, however, is seen to saturate at a constant value of ~ 0.25 emu/cm^2. One possible explanation would be a background anti-ferromagnetic phase. However, preliminary structural analysis does not indicate extraneous crystalline phases, and does verify lattice substitution of the Mn for In. Samples annealed in Sb overpressure showed no clear indications of ferromagnetism.

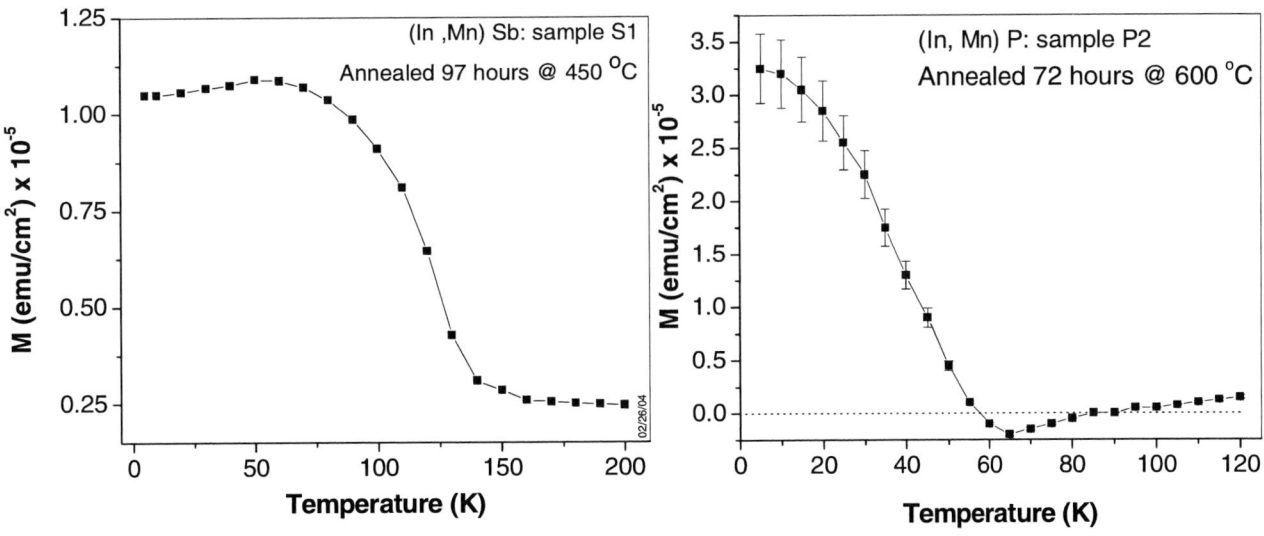

Figure 1 Magnetization plotted as a function of Temperature for (a) InSb co-annealed with Mn for 97 hours at 450 °C, T_c~130 K (b) InP co-annealed with Mn for 72 hours at 600 °C, T_c~ 55 K.

Magnetization of InP co-annealed with Mn (Figure 1b) is measured with a ferro-magnetic Curie temperature of around 55 K, comparable to that obtained by ion-implantation and subsequent annealing[19]. As with InSb, behavior above this temperature is consistent with a non-ferromagnetic phase, not seen by diffractometry. P overpressure and Cr co-doping have not produced convincing evidence of ferromagnetism in InP.

In the case of GaAs, co-annealed in the presence of both Mn and Cr, we observe a clear ferrimagnetic behavior, with a Néel temperature (T_N) at ~ 270 K. In this case, however, additional magnetic phases were observed (Figure 2) which could possibly account for the magnetic behavior. The results above are symptomatic of problems with the characterization of magnetic semiconductors where the x-ray diffraction technique is not satisfactory. We are currently working on using magneto-optic techniques to distinguish magnetic effects intrinsic to a magnetic semiconductor from those due to the formation of extraneous phases, which are not often distinguishable in structural analysis.

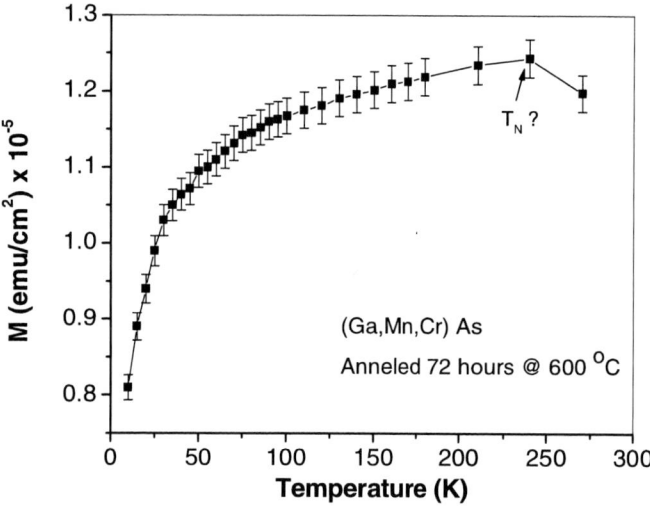

Figure 2 Possible ferrimagnetism by co-annealing GaAs with Mn and Cr. These results present the intriguing possibility of magneto-optic recording in these materials.

Structural Analysis: X-Ray Diffraction

In the case of GaAs annealed in the presence of Mn and Cr, corresponding to Figure 2 above, the x-ray diffraction (XRD) pattern indicates the presence of extraneous phases, such as MnAs (see Figure 3), which could have contributed to the observed ferrimagnetic behavior.

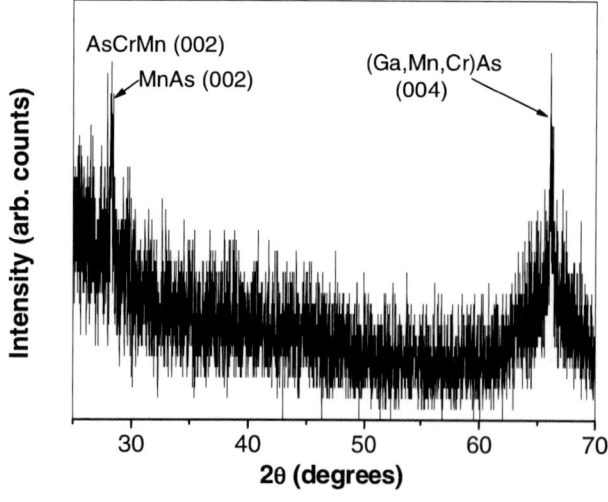

Figure 3 X-ray diffraction spectrum of GaAs co-annealed with Mn and Cr. In addition to the main peak at 2θ ~ 66.3°, orthorhombic MnAs and CrMnAs are also observed, the effects of which have to be deconvoluted to verify the incorporation of Mn and Cr into the III-V lattice.

On the other hand, we do not observe any extrinsic crystalline phases for InSb or InP annealed with Mn. However, the precipitation of metallic In at the surface was seen in some cases. This is a common problem with InP and InSb processing and could be mitigated by use of a capping layer of InP or increasing the anion vapor pressure. We estimate the concentration of Mn by analyzing the peak shift due to lattice strain (the strain due to quenched-in vacancies was estimated to be $<10^{-5}$).

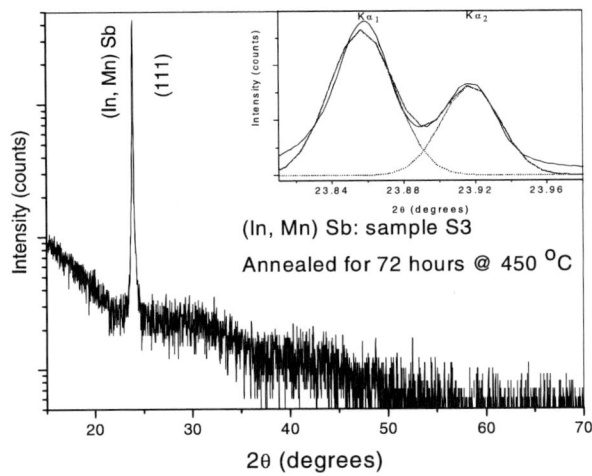

Figure 4 X-ray diffractogram of InSb co-annealed with Mn. The inset shows a Gaussian fit (dotted) of the (In, Mn) Sb (111) peak (solid), used to estimate the extent of Mn substitution into the InSb lattice.

Preliminary analysis of peak shift data suggest lattice strain up to about 0.1%, corresponding to Mn substitution for In, in InSb, of up to 5.5%. This provides proof of principle for the application of controlled ambient annealing to synthesis of magnetic semiconductors. A similar analysis of doping in InP and GaAs is in progress.

CONCLUSIONS

We propose the use of alternate, economical, controlled ambient furnace annealing procedures as a new synthesis technique for rapidly developing new magnetic semiconductors. The annealing treatments are hypothesized to increase the amount of Mn incorporation and reduce the effect of Mn interstitials, both of which are important in increasing the ferromagnetic Curie temperature (T_c). By annealing non-magnetic InSb, InP and GaAs in the presence of Mn (and Cr in the case of GaAs) we have successfully synthesized new magnetic semiconductors. The controlled ambient ampoule annealing methods being explored in this paper present an alternative way of substantially increasing T_c. Previous studies of InP: Mn prepared by ion implantation and annealing have shown a T_c of 90 K[19], comparable to our result of 55K. By contrast, InSb: Mn prepared by MBE exhibited a T_c of 8.5 K[20], but we have obtained results pointing to Curie temperatures of 130 K in InSb, with Mn additions of up to 5.5%. The possible occurrence of ferrimagnetism in Mn and Cr incorporated GaAs was also observed through SQUID magnetometry. Anion overpressure was not found to increase T_c, perhaps because antisite defects reduce the concentration of alignment-mediating holes; work is in progress to test this hypothesis. X-ray diffraction was used to determine phase homogeneity. Further study, using magneto-optic characterization and transport measurements, is expected to yield additional insight into the proposed means for increasing the incorporation of Mn, the influence of Mn segregation and the occurrence of metastable phases, all of which are major problems in the field today.

ACKNOWLEDGMENTS

We acknowledge the use of facilities at the W.M. Keck Center for Interface Materials Science at UCSD.

REFERENCES

1. H. Ohno, *Science,* **281**, 951 (1998).
2. Y. D. Park, *et al*, *Science*, **295,** 651 (2002).
3. S. Datta, and B. Das, *Appl. Phys. Lett.,* **56**, 665 (1990).
4. J.F. Gregg *et al., J. Phys. D,* **35**, R121, (2002).
5. R. de Sousa and S. Das Sarma, *Phys. Rev. B,* **67**, 033301 (2003)
6. D. Chiba, M. Yamanouchi, F. Matsukura, F., and H. Ohno, *Science* **301**, 943 (2003).
7. T. Dietl *et al., Science,* **287,** 1019 (2000).
8. M. Takahashi, and K. Kubo, *LANL archives,* arxiv:cond-mat/0204124 (2002).
9. W. Walukiewicz, *Appl. Phys. Lett.,* **54**, 2094 (1989).
10. R. Kawakami *et al, Appl. Phys. Lett.,* **77**, 2379 (2000).
11. J. Blinowski, and P. Kacman, *Phys. Rev. B,* **67**, 121204-1 (2003).
12. K.M. Yu, *et al., Phys. Rev. B,* **65**, 201303 (R), (2002).
13. D. Chiba *et al, Appl. Phys. Lett.,* **82**, 3020 (2003).
14. S.J. Potashnik *et al, Appl. Phys. Lett.,* **79**, 1495 (2001).
15. C.H. Wu and K.C. Hsieh, *J. Appl. Phys.,* **72,** 5642 (1992).
16. S. Yu, T.Y. Tan, and U. Gosele, *J. Appl. Phys.,* **70,** 4827 (1991).
17. S.J.-Han *et al, Appl. Phys. Lett.,* **83**, 920 (2003).
18. M. Linnarsson, *et al., Phys. Rev. B,* **55**, 6938, (1997).
19. Y. Shoon *et al, Appl. Phys. Lett.,* **84**, 2310 (2004).
20. T. Wojtowicz *et al, Appl. Phys. Lett.,* **82**, 4316 (2003).
21. A.R. West, *Solid State Chemistry and its Applications,* John Wiley, New York, (1992).
22. E. F. Schubert, *Doping in III-V Semiconductors,* Cambridge Univ. Press, (1993).
23. M.S. Seltzer, *J. Phys. Chem. Solids,* **26**, 243, (1965).
24. http://www.webelements.com/
25. S. Das Sarma, E.H. Hwang, and A. Kaminski, *Phys. Rev. B,* **67**, 155201, (2003).

Mat. Res. Soc. Symp. Proc. Vol. 825E © 2004 Materials Research Society

Optical properties of transition metal (Mn, Fe and V) doped Zinc Oxide ceramics and thin films

Neha Awasthi, P. Bhattacharya and R.S. Katiyar
Department of Physics, University of Puerto Rico, San Juan PR-00931.

ABSTRACT

Transition metal doped (Mn, Fe and V) ZnO ceramics and their thin films were prepared by pulsed laser deposition on glass substrates. The ceramic targets did not show any additional phase formation from XRD measurement, except for high Fe concentrations. Optical absorption showed sub-band gap absorption for Mn doping. The band gap was shifted by 0.06 eV for V doped ZnO as concentration was increased to 10%. Micro Raman spectra showed some defect induced modes for all the transition metals doped ZnO ceramics. In V doped ZnO ceramics there was two phonon induced resonance Raman scattering with increase in dopant concentrations. Raman spectra for thin films did not show any significant additional modes for Mn and Fe. However V doped ZnO thin films showed an additional mode for concentration $\geq$5 %.

INTRODUCTION

Dilute magnetic semiconductors (DMS) involve charge and spin degrees of freedom in a single material and are extensively investigated for potential applications in spintronics[1]. This field exists between the magnetism and electronics of semiconductors. Among these, II-IV semiconductors have an advantage, where the concentration of charge and spin can be controlled independently by changing the concentrations of dopant elements, injecting carriers and transition metal (TM) element respectively[2]. One of the extensively studied materials is p-type $Ga_{1-x}Mn_xAs$ for spintronic related device applications. The Curie temperature for this material is much below room temperature[1]. To achieve practical device application it is important to increase T_c above room temperature. Several theoretical predictions have raised the possibility of ferromagnetism above room temperature in 3d TM doped ZnO[3]. Zinc Oxide (ZnO) is a wide band gap semiconductor (3.3 eV) with a large exciton binding energy (60meV) and has potential for room temperature ferromagnetism. With large electron mass, $0.3m_e$ (m_e: bare electron mass), it is expected to exhibit a strong magnetic interaction between the mobile carriers and the localized magnetic ions. It is a good choice for forming transparent ferromagnetic material[4]. In this work, we investigated the optical properties of TM doped ZnO. The ceramic targets and thin films of TM doped ZnO have been characterized using X-ray diffraction, optical absorption and micro Raman spectroscopy.

EXPERIMENT

ZnO thin films with different concentrations (1,3,5,10 %) of Mn, Fe and V were deposited on glass slides using pulsed laser ablation technique. Targets for laser ablation were prepared using conventional powder processing method with high purity ZnO, Mn_2O_3, Fe_2O_3 and V_2O_5 powders. The constituent powders were taken in stoichiometric amounts and ball milled for 24 h. Dry powders were calcined at 800 °C for 4 hours. Polyvinyl alcohol (5%) was used as binder and steric acid as lubricant to prepare the pellets in a hydraulic press. Pellets were sintered at 600 °C for 2 h at 3° C/min and then at 1000 °C for 4 h at 6 °C/min.

Transparent glass slides were used as substrate after cleaning in warm soap solution, distilled water and ultrasonic in water for 3 minutes and finally dried in nitrogen flow. Excimer laser (KrF, 248nm) with laser energy 2.5-3 J/cm^2 was used to deposit the films. The deposition chamber was evacuated to 1×10^{-6} Torr background pressure. The substrate temperature was maintained at 400 °C. Oxygen was introduced keeping the pressure at 4mTorr. The target to substrate distance was maintained at 5 cm. Siemens D5000 x-ray diffractometer (XRD) with CuKα radiation was used to study the phase formation of ZnO and ZnO doped ceramics and thin films. Optical band gap measurements were done using UV-VIS spectrometer (Perkin Elmer model RS-2) in the range of 300-800 nm. The micro-Raman measurements were performed in the backscattering geometry using Jobin-Yvon T64000 Triple-mate instrument. Radiation of 514.5 nm from a Coherent Argon ion laser was focused to ~ 1 μm in diameter on the samples. A charge coupled device (CCD) system was used to collect and process the scattered light.

RESULTS AND DISCUSSION

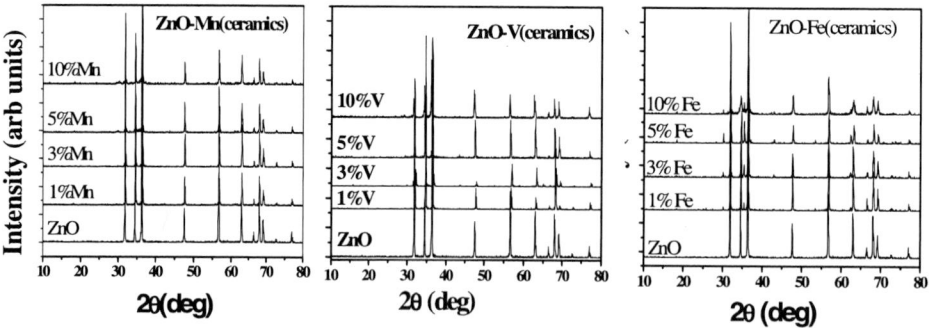

Fig. 1. XRD pattern for TM doped ZnO ceramics

The structural information about the transition metal doped ceramics and thin films were obtained using XRD. Figure 1 shows the XRD patterns for ceramic targets. The XRD pattern for Mn doped ZnO and V doped ZnO ceramics show all the peaks corresponding to the standard ZnO powder diffraction pattern with wurtzite structure. There was no additional peak corresponding to secondary phase formation in Mn and V doped compounds. Therefore, the addition of Mn and V did not cause any structural change in ZnO. Incase of V the peaks showed a shift with increase in concentration towards lower angle. However, for Fe doped ZnO additional peaks were observed at 30.078°, 35.388°, 42.867° and 62.300°. These

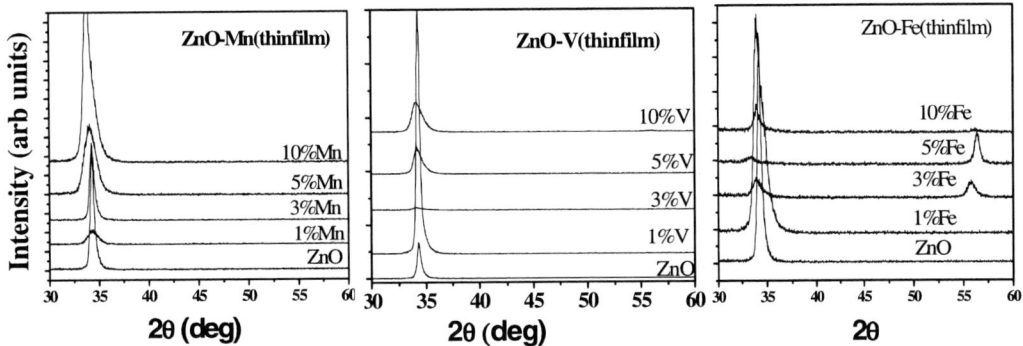

Fig. 2. XRD pattern for TM doped ZnO thin films on glass substrates

additional peaks nearly correspond to zinc ferrite and ferric oxide.

XRD pattern for thin films for Mn and V doped ZnO show only peaks corresponding to (002) at 34.5° –34.6° as in fig. 2. The films were highly c-axis oriented without any segragation of Mn or V. However, for increase in Fe concentration (>1%), an additional peak was observed at 55.8°-56.45°. This corresponds to iron related oxides. There is some additional phase formation with higher Fe concentrations (3 %), which is in consistence with the targets. The peak position for (002) peak of ZnO was shifted towards lower 2θ values with doping of

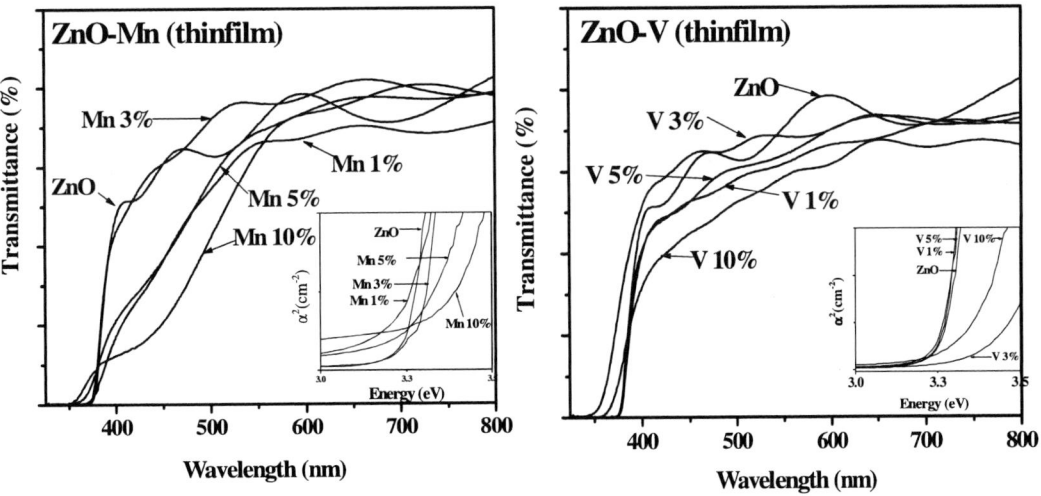

Fig. 3. Optical transmittance spectra for Mn and V doped ZnO thin films.
Inset: absorption band edge

3d TM elements. This peak shift was from 34.35° to 34.05° for dopant concentrations of 1-10 %. The c-axis value did not increase significantly with increase in concentration.

The effect of TM doping on the optical absorption of ZnO was measured using UV-VIS spectrometer. Pure ZnO films deposited on glass showed good transmission in the visible range from 400-800 nm. The band gap of ZnO and TM doped ZnO thin films was calculated using $\alpha \propto (E_g - h\nu)^{1/2}$ where α = optical absorption coefficient, E_g is the band gap and $h\nu$ incident photon energy. The value of α was calculated using the transmission data.

Mn doped ZnO showed clear sub-band gap absorption as seen from the change in transmittance(fig 3). The mechanism for this observation has been reported earlier[5]. The band edge does not change significantly from pure ZnO (3.26 eV). For V doped ZnO (fig.3) transmittance is reduced with increase in concentration. The band edge does not change for

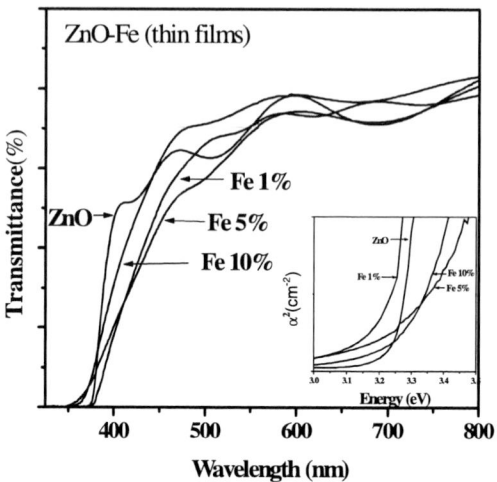

Fig. 4. Optical transmittance spectra for Fe doped ZnO thin films. Inset: absorption band edge

concentrations ≤ 5%. For 10% V concentration there is a slight shift in the band gap (3.32 eV). For Fe doped ZnO, (fig. 4) the band gap decreases for initial concentration (1%) from 3.26 eV (pure ZnO) to 3.23 eV. There is no sharp band edge observed for higher Fe concentrations.

Raman spectra of TM doped ZnO ceramic targets and thin films were taken at

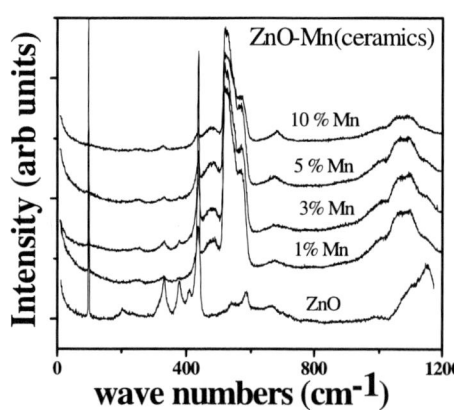 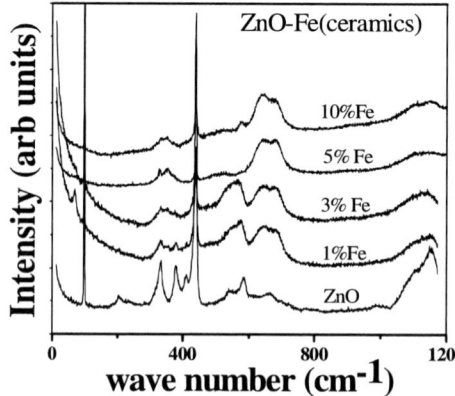

Fig.5. Raman spectra for Mn and Fe doped ZnO ceramics

room temperature. For pure ZnO target all of the modes observed correspond to wurtzite ZnO crystal structure. All atoms of the ZnO primitive cell occupy the 2b sites of symmetry C_{3v}. The wurtzite ZnO belongs to C_{6v}^4 space group and group theory predicts the existence of the following optical modes at the Γ point of the Brillouin Zone: $\Gamma_{opt} = A_1 + 2B_1 + E_1 + 2E_2$. The A_1 and E_1 modes are both infrared active and Raman active and therefore these can have longitudinal and transverse frequencies (LO and TO).

Raman spectra for Mn doped ZnO show E_2 modes of ZnO at 98.8 cm^{-1} and 438.0cm^{-1} where the intensity of higher mode reduces with increase in Mn concentration. The A_1T (378.7 cm^{-1}) and E_1T (410.8 cm^{-1}) modes are weakly observed with 1% and higher Mn doping. There is an intense second order mode at 332.6cm^{-1} which falls in intensity with increase in Mn concentration. There are strong modes at 523.5cm^{-1} (second order

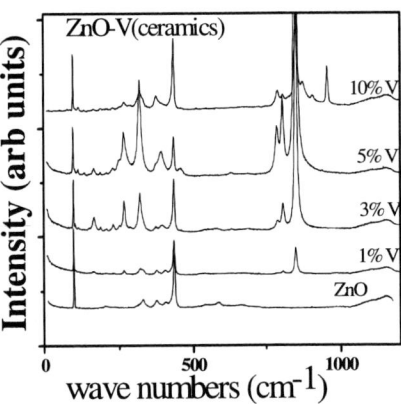

Fig. 6. Raman spectra of V doped ZnO ceramics

mode) and 568.0 cm^{-1} (E_1 LO) for all doping concentrations. These LO phonons get enhanced with increase in Mn doping concentration. The intensity of the multiphonon band at ~1100cm^{-1} is also affected by doping. It shifts to lower frequency with increase in dopant concentration. This has been attributed in Froehlich interaction of electrons and the longitudinal field produced by the LO phonons. This feature was observed for all three TM doped ceramics. Raman spectra for Fe doped ZnO show the same E_2 modes as in ZnO. There is an enhancement in the LO modes with increase in concentration. The TO modes show reduced intensity with increase in concentration up to 3%. An additional mode appears at 665.8cm^{-1} that is not yet understood. Raman spectra of V doped ZnO ceramics also show sharp and strong resonance enhanced Raman scattering at 377.1cm^{-1}, 395.8 cm^{-1}, 438.2 cm^{-1}, 788.2 cm^{-1}, 808.3cm^{-1}, 850.9cm^{-1}. The additional modes at 876.2cm^{-1}, 909.1cm^{-1} and 957.1cm^{-1} are attributed to V_xO_y modes, oxides for higher V concentration.

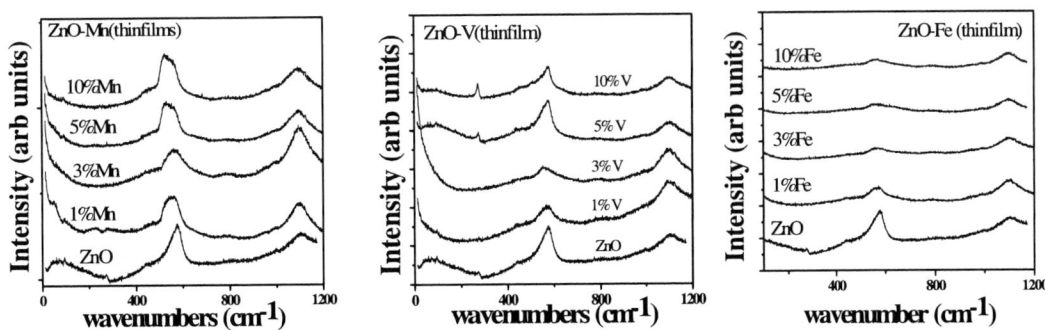

Fig. 7. Raman spectra of TM doped ZnO thin films on glass substrates

Raman spectra of Mn and Fe doped ZnO thin films do not show any additional modes. However V doped thin films show an additional mode at 275.1cm^{-1} for 5%

and 10% concentrations. Further study is in progress to investigate these additional modes for V doped thin films and ceramics.

CONCLUSION

Transition metal (Mn, V and Fe) doped thin films of ZnO have been fabricated by pulsed laser ablation technique. The ceramic targets did not show any additional phase formation from XRD measurement, except for high ($\geq 3\%$) Fe concentrations. Fe did not form a good solid solution with ZnO for high concentrations. Optical absorption shows sub-band gap absorption with increase of Mn doping. For V doped ZnO there was shift of 0.06eV in band gap as concentration is increased to 10%. Doping of Fe did not show a sharp band edge for concentrations $\geq 1\%$. Micro Raman spectra showed some additional defect induced modes for TM doped ZnO ceramics. Phonon induced resonance Raman scattering, with increase in concentration was observed in V doped ZnO ceramics. Raman spectra for thin films did not show any significant additional modes for Mn and Fe. However V doped ZnO thin films showed an additional mode for concentration ≥ 5 %. Work is in progress to understand the origin of additional Raman modes for TM doped ZnO.

ACKNOWLEDGEMENTS

The authors acknowledge partial financial supports from DE-FG-02-01ER45868 and NASA#NCC3-1034 and DAAD 19-03-1-0084 grants.

REFERENCES

1. H. Ohno, *Science* **281**, 951 (1998).
2. J. K. Furdyna, *J. Appl. Phys*, **64**, R29 (1998).
3. T. Dietl et al., *Science* **287**, 1019 (2000).
4. K. Ueda et al., *Appl. Phys. Lett* **79**, 988 (2001).
5. T. Fukumara et al., *Appl. Phys. Lett* **75**, 3366 (1999).
6. J.M. Calleja and M. Cardona, *Phy Rev B* **16**, 3753 (1977).

Valencies of Mn impurities in ZnO

L. Petit[1], T. C. Schulthess[1], A. Svane[2], W.M. Temmerman[3], and Z. Szotek[3]

[1] *Computer Science and Mathematics Division, and Center for Computational Sciences, Oak Ridge National Laboratory, Oak Ridge, TN 37831, USA*

[2] *Institute of Physics and Astronomy, University of Aarhus, DK-8000 Aarhus C, Denmark*

[3] *Daresbury Laboratory, Daresbury, Warrington WA4 4AD, UK*

Abstract

We use the self-interaction corrected (SIC) local spin-density (LSD) approximation to investigate the groundstate valency configuration of Mn impurities in p-type ZnO. In $Zn_{1-x}Mn_xO$, we find the localized Mn^{2+} configuration to be preferred energetically. When codoping $Zn_{1-x}Mn_xO$ with N, we find that four d-states stay localized at the Mn site, while the remaining d-electron charge transfers into the hole states at the top of the valence bands. If the Mn concentration [Mn] is equal to the N concentration [N], this results in a scenario without carriers to mediate long range order. If on the other hand [N] is larger than [Mn], the N impurity band is not entirely filled, and carrier mediated ferromagnetism becomes theoretically possible.

The design of diluted magnetic semiconductors (DMS), that apart from the well known electronic properties also have incorporated spin-functionality, is expected to play a major part in the development of the next generation of electronic devices. [1] In this respect, it is important that these materials remain ferromagnetic above room temperature. In Mn doped GaAs, where ferromagnetism is well established, the Curie temperature is $T_C \simeq 160$ K. Ferromagnetism has been predicted theoretically to occur in a variety of Mn doped semiconductors, but there remains considerable disagreement as to the nature of the exchange mechanism and the magnetic order.

ZnO crystallizes in the hexagonal wurtzite structure (lattice constants a_0=3.2495 Å, and c_0=5.2069 Å). Its wide band gap, in the near UV range (3.3 eV) makes it a candidate for optoelectronic applications that rely on short wavelength light emitting diodes. As was shown by Fukumura *et al.*, [2] solubility of Mn in the ZnO matrix is relatively high (x $\leq$ 0.35). The various experimental investigations of the magnetic order in $Zn_{1-x}Mn_xO$ give contradictory results, ranging from spin glass behaviour [3] and paramagnetism, [4] to ferromagnetism at room temperature [5]. The very latest experimental study that we are aware of, finds no evidence for magnetic order, down to T=2 K. [6], and it has been suggested that the previously observed ferromagnetism is due to precipitates containing manganese oxides. [7] Codoping with N has so far revealed itself to be rather elusive. [8]

With respect to theory, the agreement is that $Zn_{1-x}Mn_xO$, without additional carriers is not ferromagnetic. According to the Zener model approach by Dietl *et al.*, [9] ferromagnetism in DMS's originates from the RKKY-like interaction between the localized transition metal moments, and delocalized hole carriers. In Mn doped ZnO, the Mn impurities provide the localized moments, but without acceptor codoping, there are no carriers to mediate the long range interaction. Bandstructure calculations, based on the LSD approximation, [10, 12, 5, 11] find the delocalized Mn-d levels to be situated in the ZnO semi-conducting gap. Without additional hole doping, the Fermi level separates a completely filled majority-spin band from a completely empty minority-spin band resulting in a spin glass state. [10].

The question as to which picture, Zener or LSD, is more appropriate to describe $Zn_{1-x}Mn_xO$ depends on the relative strength of the on-site d-d correlations which tend to localize the $5d$-electrons on each their site, and the gain in band formation energy that results from the electrons becoming delocalized over the crystal. In order to evaluate the actual d-electron groundstate configuration, one needs to be able to compare the total energies of the two scenarios. With the SIC-LSD approximation, used in the present work, [13, 14] delocalized and localized d-electrons are treated on an equal footing, by adding a contribution to the LSD energy functional, allowing each d-electron to localize. [15] Thus the localized mean field scenario and the delocalized bandstructure scenario can be compared, and the groundstate configuration of the Mn-ion can be deduced from the global energy minimum.

When doping ZnO with Mn, the Mn ions occupy the Zn site without changing the wurtzite structure. [2] For our initial calculations, we consider a (2x2x2) supercell consisting of 8 ZnO unit cells, with a single Zn substituted by Mn. Changing the valency configuration of the Mn ion in $Zn_{1-x}Mn_xO$, from localized Mn^{2+} to delocalized Mn^{7+}, enables us to study the resulting changes in electronic structure, and to compare the corresponding total energies. The main graph in Fig. 1 shows the density of states (DOS), as obtained in the localized scenario, i.e. keeping five d-electrons on the Mn ion localized. The inset in the same figure shows the DOS, as obtained in the LSD approximation, i.e. with the Mn d electrons delocalized. The main difference

G2.9.3

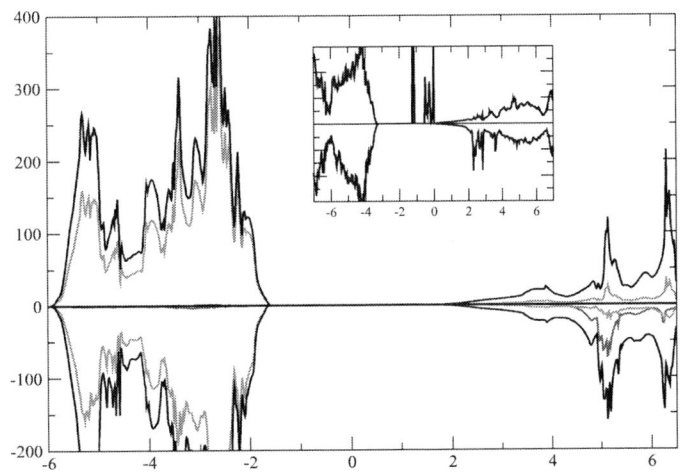

Figure 1: Total DOS, in states per eV, of $Zn_{1-x}Mn_xO$, with Mn respectively in the divalent Mn^{2+} configuration (main graph), and the delocalized Mn^{7+} configuration (inset). The black, purple, and green lines represent the total, d-projected, and p-projected densities of states respectively. The energies are in eV, with the Fermi energy at zero.

between the two plots, is that the band features, which in the delocalized scenario are situated just below the host conduction edge, have disappeared in the SI corrected scenario. The LSD picture is in good agreement with the previously mentioned *ab-initio* calculations. However, comparing the total energy for both scenarios, we find that the localized Mn^{2+} configuration is energetically favoured over the LSD scenario by more than 3 eV. Localizing the Mn $3d$'s on each their site thus results in an overall gain in SIC energy that far outweighs any corresponding loss in hybridization energy. The Mn^{2+} valency groundstate configuration is in agreement with experimental observation, [16] and would seem to support the Zener model rather than the band picture.

As was stated earlier, theory predicts ferromagnetism to occur in Mn-doped ZnO, only if additional hole carriers are introduced. In the Zener model the holes guarantee the communication between the localized sites, whilst according to the bandstructure calculations, due to the hole states, the majority spin-band is only partially filled, and the ferromagnetic double-exchange becomes strong enough to overcome the anti-ferromagnetic super-exchange. [10] We p-dope $Zn_{15/16}Mn_{1/16}O$ by substituting a single O atom with N. Gradually delocalizing the Mn d-electrons, we find that the global

15

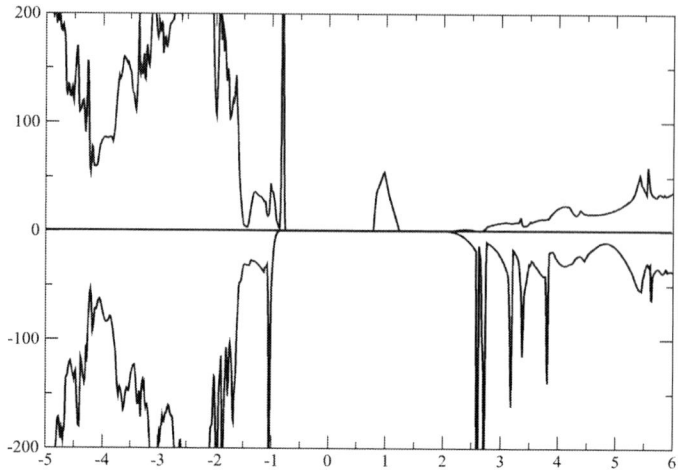

Figure 2: Total DOS (in states per eV) of $Zn_{1-x-y}Mn_xN_yO$ in the trivalent Mn^{3+} configuration. The energies are in eV, with the Fermi energy at zero .

energy minimum is now obtained in the trivalent Mn^{3+} with four localized d-electrons. The resulting DOS for the groundstate configuration is shown in Fig. 2. The narrow, exchange-split peak, at the top of the valence band originates from the p-orbitals of the N-impurity. Compared to the O which it substitutes, N has one less occupied p-orbital, and in the scenario with 5 localized Mn d-electrons, this impurity peak would remain partially unoccupied. In the actual groundstate configuration, one of the otherwise localized d-states gains sufficient energy, through hybridization and charge transfer to overcome the corresponding loss in SIC energy, and prefers to delocalize, which results in the valence band being completely filled, and the Fermi level being situated in the band gap, below the delocalized Mn-d peak. From the SIC-LSD calculations it thus follows that neither the itinerant nor the localized scenarios represent the groundstate of $Zn_{15/16}Mn_{1/16}O_{15/16}N_{1/16}$ and that there are no hole carriers in the actual trivalent groundstate. Alternatively to N doping, we also considered P and C as possible hole donors, but also in this case the Mn^{2+} groundstate configuration is not realized.

The question arises to what degree the delocalization depends on the proximity of the Mn and N dopants. In order to determine the effect of the Mn-N distance on the preferred valency configuration of the Mn ion, we calculated the total energies for different scenarios, where the N impurity is substituting an O atom, situated at increasing distances from the Mn impurity. It turns out, that the trivalent configuration remains energetically most favourable, in all three cases, however we notice that the energy difference between the divalent and trivalent configuration, $E(Mn^{2+})$-$E(Mn^{3+})$, decreases from the nearest neighbour scenario to the furthest away neighbour scenario. Overall, the lowest total energy is obtained for Mn and N substituting two neigbouring Zn and O sites, which might be an indication that the Mn ions will tend to cluster around the N impurity.

Even though the codoping with N readily removes one electron from the otherwise localized Mn-d manifold, the resulting oxidation state is stable with respect to an increase in N concentration. Comparing the total energies of $Zn_{15/16}Mn_{1/16}O_{14/16}N_{2/16}$ for respectively the Mn^{2+}, Mn^{3+}, and Mn^{4+} configurations, we find the Mn^{3+} configuration with four localized d-electrons to be energetically most favourable. In this scenario, the Fermi level is situated in the now only partially filled, N impurity band at the top of the valence band, and the coexistence of both localized spins from the Mn^{3+} and hole states in the valence bands, makes ferromagnetism, as proposed in the Zener model, theoretically possible.

In conclusion, we have studied the electronic structure of Mn impurities in ZnO. With no additional carriers, we find the localized Mn^{2+} to be energetically most favourable. When codoping with N, the Mn^{3+} configuration is realized, and ferromagnetism mediated by hole carriers becomes possible if the N concentration is larger than the Mn concentration.

This work was supported in part by the Defense Advanced Research Project Agency and by the Division of Materials Science and Engineering, US Department of Energy. Oak Ridge National Laboratory is managed by UT-Batelle, LLC, for the US Department of energy under Contract No. DE-AC05-00OR22725.

References

[1] S. A. Wolf, D. D. Awschalom, R. A. Buhrman, J. M. Daughton, S. von Molnar̓, M. L. Roukes, A. Y. Chtchelkanova, and D. M. Treger, Science **294**, 1488 (2001).

[2] T. Fukumura, Zhengwu Jin, A. Ohtomo, H. Koinuma, and M. Kawasaki, Appl. Phys. Lett. **75**, 3366 (1999).

[3] T. Fukumura, Zhengwu Jin, M. Kawasaki, T. Shono, T. Hasegawa, S. Koshihara, and H. Koinuma, Appl. Phys. Lett. **78**, 958 (2001).

[4] A. Tiwari, C. Jin, A. Kvit, D. Kumar, J. F. Muth, and J. Narayan, Solid State Commun. **121**, 371 (2002).

[5] P. Sharma, A. Gupta, K. V. Rao, F. J. Owens, R. Sharma, R. Ahuja, J. M. O. Guillen, B. Johansson, and G. A. Gehring,Nature materials **2**, 673 (2003).

[6] G. Lawes, A. P. Ramirez, A. S. Risbud, and Ram Seshadri, cond-mat/0403196.

[7] Y. M. Kim, M. Yoon, I. -W. Park, Y. J. Park, and Jong H. Lyou, Solid State Commun. **129**, 175 (2004).

[8] M. Joseph, H. Tabata, and T. Kawai, Jpn. J. Appl. Phys. **38**, L1205 (1999).

[9] T. Dietl, H. Ohno, F. Matsukura, J. Cibert, and D. Ferrand, Science **287**, 1019 (2000).

[10] K. Sato, and H. Katayama-Yoshida, Semicond. Sci. Technol. **17**, 367 (2002).

[11] N. A. Spaldin, Phys. Rev. B **69**, 125201 (2004).

[12] Yu. Uspenskii, E. Kulatov, H. Mariette, H. Nakayama, and H. Ohta, JMMM **258-259**, 248 (2003).

[13] A. Svane, Phys. Rev. B **53**, 4275 (1996).

[14] W. M. Temmerman, A. Svane, Z. Szotek and H. Winter, in *Electronic Density Functional Theory: Recent Progress and New Directions*, eited by J. F. Dobson, G. Vignale and M. P. Das (Plenum, New York, 1998), p. 327.

[15] A. Zunger, J. P.Perdew, and G. L.Oliver, Solid State Commun. **34**, 933 (1980).

[16] P. B. Dorain, Phys. Rev. **112**, 1058 (1985).

Model for Spin Injection into Conjugated Organic Semiconductors

P. Paul Ruden
Department of Electrical and Computer Engineering, University of Minnesota,
Minneapolis, MN 55455, USA
Darryl L. Smith
Los Alamos National Laboratory
Los Alamos, NM 87545, USA

ABSTRACT

We present a theoretical model to describe electrical spin injection from a ferromagnetic metal contact into a conjugated organic semiconductor. To achieve significant spin current, the organic semiconductor must be driven far out of local thermal equilibrium by an electric current. Effective spin injection therefore requires that equilibration between the conjugated organic semiconductor and the metallic contact be suppressed by an energy barrier to injection that may be due either to a large Schottky barrier or to an insulating tunnel barrier. The results are compared with simulations for a silicon based device structure. Detection of the injected spin current in the organic semiconductor is also addressed.

INTRODUCTION

Extensive recent electronic materials research has focused on physical phenomena that involve the spin degree of freedom of mobile charge carriers. [1] In part, this interest is motivated by the prospect of using spin, in addition to charge, as an information carrying physical quantity in electronic devices. Ferromagnetic metals provide a convenient source of spin-polarized electrons for injection, and they can serve as an electrical means to detect a spin-polarized current at a device terminal. Several experiments have successfully demonstrated spin injection from ferromagnetic metals into conventional (inorganic) semiconductors. [2, 3, 4] Spin polarization lifetimes in inorganic semiconductors are primarily limited by the spin-orbit interaction.

Very recently spin injection into organic semiconductors and electrical detection of the spin current have also been reported. [5, 6] Since these materials are composed of light elements – principally carbon and hydrogen – the spin-orbit interaction is small and spin lifetimes of charge carriers are expected to be comparatively long. Organic semiconductors are π-conjugated materials, either small molecules or polymers, with energy gaps ranging from about 1.5eV to 3.5eV. They are usually undoped, but charge carriers may be injected from metallic contacts. Electronic devices can be fabricated from disordered thin films, in which electrical conduction results from carrier hopping between localized sites and carrier mobilities are therefore quite low. At metal contacts to organic semiconductors, the Schottky energy barrier usually scales directly with the contact metal workfunction. [7]

Early calculations of spin injection from ferromagnetic metals into nonmagnetic semiconductors showed that the large difference in conductivity of the two materials inhibits the creation of such a non-equilibrium electron spin distribution, and thus makes efficient spin injection difficult. [8] In subsequent work on electron injection from ferromagnetic metals into inorganic semiconductors it was shown that a spin selective interface resistance, such as is provided by a tunneling barrier, can be effective for achieving efficient spin polarization of the injected current.[9, 10] The tunnel barrier can be formed either by adding a thin insulating layer between the metal contact and the semiconductor or by using an interface doping profile to engineer a tunneling region in the depletion region of the Schottky contact.[11] Tunneling through a potential barrier can be spin selective because the barrier transmission coefficient depends not only on the barrier energy profile, but also on the wavefunction of the tunneling particle in the contact regions. The wavefunctions are different for spin up and spin down electrons at the Fermi surface of a ferromagnet. Organic semiconductors may be particularly attractive for the fabrication of ordered tunnel barriers because self-assembly techniques can be used to grow ordered monolayers at a contact.[12]

If the charge carrier injecting and extracting contacts are both ferromagnetic, their relative direction of magnetization changes the magnitude of the charge current for a fixed applied voltage. We investigate the dependence of the charge current on the relative direction of magnetization of the two contacts at a fixed bias in an organic diode structure with two ferromagnetic contacts.

MODEL DISCUSSION

We consider electrical spin injection in an organic semiconductor device structure, consisting of a thin (100nm) organic semiconductor film, sandwiched between two ferromagnetic metal contacts. We specifically take a case in which the workfunctions of the two metal contacts are chosen so that electrons are injected efficiently and the hole concentration is always small. A device model previously developed[13] is generalized to include spin injection and transport.[14] The coupled Poisson and drift-diffusion equations are solved and boundary conditions for the drift-diffusion equation are determined by interfacial carrier injection. We include the possibility of a tunnel barrier at the contact.

It is assumed that there is no strong spin scattering as electrons traverse the interfacial layer between the metal and the semiconductor so that both charge and spin currents are continuous across this interface. We characterize the interfacial layers by interface resistances $R_{\uparrow,\downarrow}$ for spin up and spin down electrons, respectively, and we assume that $R_\uparrow < R_\downarrow$ if the ferromagnet is polarized such that $\sigma_\uparrow > \sigma_\downarrow$. For the numerical calculations we take $R_\uparrow = (1/2)R_\downarrow$.

For each spin type, the injected current, $J_{inj;\uparrow,\downarrow}$, is the sum of a thermionic emission current, an interface recombination current, and a tunneling current, which enables electron tunneling through the potential barrier of the Schottky contact:

$$J_{inj;\uparrow,\downarrow} = (1/2)A^*T^2 \exp(-\Phi_{B;\uparrow,\downarrow}/kT) - A^*T^2 n_{\uparrow,\downarrow}(0^+)/n_0 + J_{tun;\uparrow,\downarrow} \qquad (1)$$

Here, A^* designates Richardson's constant, T the temperature, k Boltzmann's constant, and $\Phi_{B;\uparrow,\downarrow}$ the Schottky barrier height, which depends on spin because the quasi Fermi levels in the contact are different for the two spin directions. The electron densities for spin up and spin down carriers in the semiconductor at the interface are denoted by $n_{\uparrow,\downarrow}(0^+)$, and $J_{tun;\uparrow,\downarrow}$ is the spin-dependent tunneling current through the Schottky potential barrier. Consistent with the spin selectivity of the interface resistance, $J_{tun;\uparrow} > J_{tun;\downarrow}$ if $\sigma_\uparrow > \sigma_\downarrow$. In the numerical calculations we take spin up electrons to have twice the transmission probability of spin down electrons. The first two terms, but not the third, on the right side of equation (1) are related by detailed balance. Image charge induced Schottky barrier lowering, which is comparatively strong in the organic semiconductors due to their small dielectric constants, is included in the determination of the value of $\Phi_{B;\uparrow,\downarrow}$ for given local electric fields at the interface.

The current boundary condition at the electron extracting contact is formulated in direct analogy to that for the injecting contact. Because of the direction of the electric field there is no tunneling through the Schottky barrier region and no image force lowering of the Schottky barrier at the extracting contact.

In the organic semiconductor the charge current density, $j = j_\uparrow + j_\downarrow$, satisfies the continuity equation, and the steady state spin current density, $j_s = j_\uparrow - j_\downarrow$, and the spin polarized electron density, $n_s = n_\uparrow - n_\downarrow$, are related by a continuity equation of the form,

$$dj_s / dx - 2en_s / \tau_s = 0, \tag{2}$$

where spin relaxation is described by a time constant, τ_s. The electron mobility of the organic semiconductor is taken to have the Poole-Frenkel form, $\mu(F) = \mu_0 \exp((|F|/F_0)^{1/2})$, where F is the electric field and μ_0 and F_0 are material parameters.

RESULTS AND DISCUSSION

We first consider a structure consisting of an organic semiconductor sandwiched between two ferromagnetic contacts with parallel polarization ($\sigma_\uparrow > \sigma_\downarrow$) and without interfacial tunnel barriers ($R_\uparrow = R_\downarrow = 0$). The metal forms a Schottky contact to the organic semiconductor. Fig. 1 shows the calculated injected charge and spin current densities near the injecting contact for a Schottky barrier height of 0.8eV. The spin polarization of the current density, for large charge current density is appreciable. The greater spin polarization at high voltages is due to spin selective tunneling through the Schottky barrier region and at high voltages j_s/j approaches the fractional spin polarization of the tunneling current. The tunneling current increases more rapidly with increasing electric field near the junction than does the thermionic emission current. We plot the magnitudes of the individual contributions to the injection current arising from thermionic emission, interface recombination, and tunneling, together with the device current in Fig. 2. At low bias (and for small Schottky barrier height even at large bias), the net injection current is primarily determined by a combination of thermionic emission and interface recombination, which is nearly equal in magnitude, as is shown in Fig. 2. When electrical injection is dominated by processes related by time reversal, the electron populations on the two sides of the interface are near quasi-equilibrium because of detailed balance. Due to the high electron density in the metal contact, the spin population in the contact cannot be driven far out of local thermal equilibrium by practical current densities. Since the organic semiconductor is in quasi-equilibrium with the contact when thermionic emission and interface recombination dominate there will not be efficient spin injection. At high bias, however, the combination of tunneling and interface recombination dominate, as is seen in Fig. 2. Tunneling and interface recombination are not related by time reversal and the rates for these two processes are not connected by detailed balance. Therefore, when tunneling and interface recombination dominate

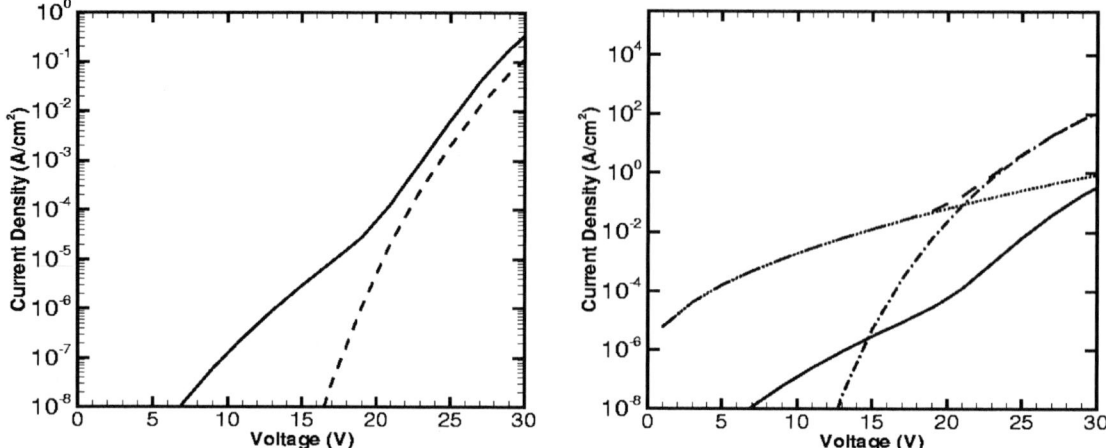

Fig. 1: Charge (solid) and Spin (dashed) current density vs. voltage.
Fig. 2: Thermionic emission (dotted), interface recombination (dashed), tunneling (dash-dotted), and total device current (solid).

injection, the electron populations on the two sides of the interface need not be near quasi-equilibrium and efficient spin injection is possible as is shown in Fig. 1.

We next consider the effect of an interfacial tunnel barrier with spin selective contact resistance. The ferromagnetic injecting and extracting contacts are both polarized such that $\sigma_\uparrow > \sigma_\downarrow$. The Schottky barrier height is 0.3eV and the contact resistances are: $R_\uparrow = 5x10^{-3}\Omega cm^2$ and $R_\downarrow = 10^{-2}\Omega cm^2$. The calculated charge and spin current densities in the organic semiconductor near the interface are displayed as a function of voltage in Fig. 3. A spin selective contact resistance breaks up the detailed balance condition and can give rise to an appreciable spin polarization of the injected current even in the case of a small Schottky barrier. Increasing the contact resistances (while keeping the ratio $R_\uparrow / R_\downarrow$ constant) monotonically increases the spin polarization at a given current density but also increases the voltage between the contacts that is required to reach a given current.

To put these results into perspective we model a comparable device structure consisting of a 100nm thick undoped silicon layer sandwiched between two ferromagnetic metal contacts. We assume a Schottky barrier height of 0.5eV and a field-dependent electron mobility of the form $\mu(F) = \mu_0 / \sqrt{1 + (F / F_0)^2}$. Results without interfacial barrier layers show that the spin current is more than six orders of magnitude smaller than the charge current in the range of electric fields that would not induce breakdown. Analyzing the injection current components shows that in this device thermionic injection dominates and is not cancelled by the interface recombination current to the same extent as in the organic structure. This may be attributed to the much larger drift velocities attained in silicon. In contrast to the organic layer the interface recombination velocity is not significantly larger than the drift velocity and, hence, electrons injected into the semiconductor from the metal have a much smaller probability of returning to the contact. Field induced barrier lowering is much less effective in this device due to the larger dielectric constant of silicon. However, for fields that are smaller than breakdown fields, the tunneling current is very small and therefore the spin-polarization of the injected current is low.

Adding an interfacial resistance of the same magnitude as used for the organic structure significantly improves the spin-polarization of the current, as is seen in Fig. 4.

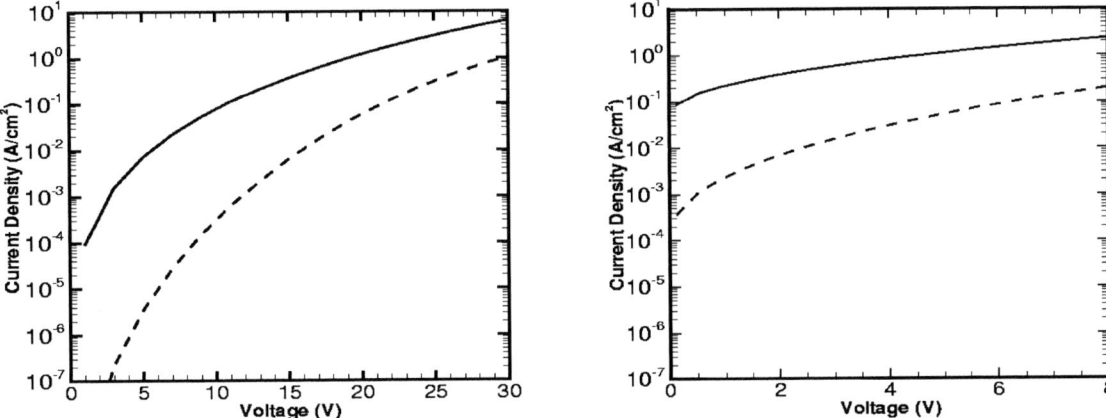

Fig. 3: Charge and spin currents for an organic structure with interface tunneling barriers.
Fig. 4: Charge and spin currents for a silicon structure with interface tunneling barriers.

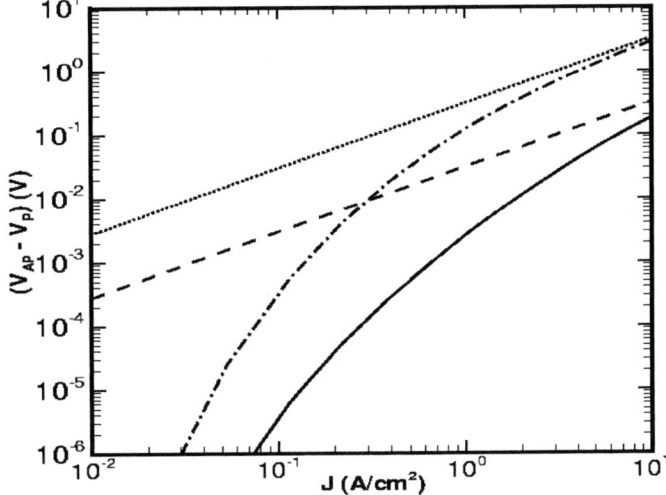

Fig. 5: Voltage difference for antiparallel (AP) parallel (P) polarized magnetic contacts as a function of the charge current. Solid and dash-dotted lines are for $\Phi_B = 0.3eV$ and dotted and dashed lines for $\Phi_B = 0.8eV$. Upper curves are for larger and lower for smaller interface resistances.

Lastly, we examine the possibility of detecting the spin current through the organic layer by measuring the difference in overall resistance of devices with parallel and anti-parallel polarization of the ferromagnetic contacts. The spin current contributes to the voltage drop across the interfacial contact layers if the latter have spin selective resistances. Fig. 5 shows the calculated voltage differences obtained assuming anti-parallel and parallel polarization of the contacts versus charge current. Results are plotted for the two Schottky barrier heights considered (0.8eV and 0.3eV), and for two different contact layer resistances ($R_\uparrow = 5x10^{-3}\Omega cm^2$

and $R_\uparrow = 5x10^{-2} \Omega cm^2$, with $R_\downarrow / R_\uparrow = 2$ in both cases). Measurable voltage differences are obtained for reasonable values of the charge current density.

CONCLUSIONS

Conjugated organic semiconductors are promising materials for the investigation of non-equilibrium electron-spin-based phenomena because these materials have excellent potential for the fabrication of novel device structures. The model for spin-polarized transport in organic semiconductors presented in this work supports the picture that spin selective tunneling is critical for the spin-polarized injection of charge carriers from a ferromagnetic metal into a semiconductor. The tunneling may arise from a strongly reverse biased Schottky contact or from a thin insulating interface layer, such as may be provided by a self-assembled monolayer on an organic material. The injected spin current is nearly constant throughout the semiconductor if the charge carrier transit time is short compared to the spin relaxation time. The voltage drop developed across the device for a given bias (charge) current is different for the cases of parallel and anti-parallel polarization of the contacts and that voltage difference should be readily detectable under conditions of acceptable currents.

ACKNOWLEDGMENT

One of us (PPR) acknowledges the assistance of J.W. Balk, partial support by NSF, 3M Inc., and the gracious hospitality of LANL. The work of DLS was supported by the Los Alamos Laboratory Directed Research and Development program.

REFERENCES

[1] See for example: "Semiconductor Spintronics and Quantum Computation," D.S. Awschalom, D. Loss, and N. Samarth, eds., Springer, Berlin (2003), and references therein.

[2] A. Isakovic, D.M. Carr, J. Strand, B.D. Schultz, C.J. Palmstrom, and P.A. Crowell, Phys. Rev. B 64, 016601 (2001).

[3] J.H. Zhu, M. Ramsteiner, H. Kostial, M. Wassermeier, H.-P. Schonherr, K.H. Ploog, Phys. Rev. Lett., 87, 016601 (2001).

[4] A.T. Hanbicki, B.T. Jonker, G. Itskos, G. Kioseoglou, and A. Petrou, Appl. Phys. Lett., 80, 1240 (2002).

[5] V. Dediu, M.Murgia, F.C. Matacotta, C. Taliani, S. Barbanera, Solid State Commun., 122, 181 (2002).

[6] Z.H. Xiong, D. Wu, Z. Valy Vardeny, and J. Shi, Nature 427, 821 (2004).

[7] I.H. Campbell and D.L. Smith, in Solid State Physics Vol. 55, edited by H. Ehrenreich and F. Spaepen, (Academic, New York, 2001).

[8] G. Schmidt, D. Ferrand, L.W. Molenkamp, A.T. Filip, B.J. van Wees, Phys. Rev. B, 62, R4790 (2000).

[9] E.I. Rashba, Phys. Rev. B, 62, R16276 (2000).

[10] D.L. Smith and R.N. Silver, Phys. Rev. B 64, 045323 (2001).

[11] J.D. Albrecht and D.L. Smith, Phys. Rev. B66, 113303 (2002).

[12] I.H. Campbell, J.D. Kress, R.L. Martin, D.L. Smith, N.N. Barashkov, and J.P. Ferraris, Appl. Phys. Lett., 71, 3528 (1997).

[13] P.S. Davids, I.H. Campbell, and D.L. Smith, J. Appl. Phys. 82, 6319 (1997).

[14] P.P. Ruden and D.L. Smith, J. Appl. Phys. 95, 4898 (2004).

Spin Effects of Low-dimensional Electron Gases Studied by Far-infrared Photoconductivity Experiments

C. -M. Hu

Institut für Angewandte Physik und Zentrum für Mikrostrukturforschung, Universität Hamburg, Jungiusstraße 11, D-20355 Hamburg, Germany

Email: hu@physnet.uni-hamburg.de

ABSTRACT

We review our recent work on spin effects in low-dimensional electron gases studied using far-infrared photoconductivity technique. We measure the spin-orbit coupling parameter α via spectroscopy by detecting the combined resonance. Detailed filling-factor dependent study shows the collective nature of this excitation, in accordance to theoretical predictions that both Kohn and Larmor theorem are broken for long-wavelength excitations that changes both the Landau and spin quantum numbers. We find that the long spin-relaxation time of a two-dimensional electron gas results in a novel bolometric spin effect, which gives rise to a substantial photo resistance change by reversing the spin polarization of electrons at the Fermi-level.

INTRODUCTION

In the classical picture, conduction electrons of a semiconductor placed in a magnetic field $\boldsymbol{B}$ feel a Lorenz force that drives the electron moving in the cyclotron orbit, while the magnetic moment of the spin feels a torque that causes the spin to precess. These motions resonantly interact with the electromagnetic radiation, with the cyclotron resonance (CR) [1] frequency

$$\omega_c = eB/m^* \tag{1}$$

and electron-spin-resonance (ESR) [2] frequency

$$\omega_z = -v_0 \omega_c \tag{2}$$

being typically in the THz and the GHz regime, respectively ($v_0 = gm^*/2m_e < 0$ for most semiconductors). They provide textbook examples [1,2] of accurate determination of the electron effective mass m^* and the Landé g factor. In the quantum mechanical picture, CR is the inter-Landau-level electric dipole transition with $\Delta N = 1$ and $\Delta S = 0$, and ESR is the inter-Zeeman-level magnetic dipole transition with $\Delta N = 0$ and $\Delta S = -1$, where N and S are the Landau and spin quantum numbers, respectively.

These are simplified pictures that neglect the nonparabolicity in narrow gap semiconductors and spin-orbit coupling in semiconductors lacking an inversion center. In both cases, coupling between the orbital and the spin motion of the electrons breaks the simple selection rules described above. In InGaAs/InAlAs heterojunctions where the structure inversion asymmetry dominates the spin-orbit coupling over the bulk inversion asymmetry [3], a combined resonance (CBR) with both the Landau and spin quantum numbers changed [4] can be excited by either the electric (**E**) or the magnetic (**H**) component of the radiation, typically with the THz frequency given by [3]

$$\omega_{CBR} = \sqrt{(\omega_c + \omega_z)^2 + (2\Delta_R/\hbar)^2} \,, \qquad (3)$$

which approaches $\omega_c + \omega_z$ only at high B fields when $\hbar\omega_c \gg 2\Delta_R/(1-\nu_0)$. Here the matrix element $\Delta_R = \alpha k_F$ depends on the spin-orbit parameter α and the Fermi wave vector k_F, which both can be controlled via a front gate [5]. The potential significance of manipulating spin via the gate is best illustrated in the classic paper of Datta and Das for a novel spintronic device [6].

Of primary importance for a two-dimensional electron gas (2DEG) is the effect of the electron-electron interaction. In the presence of a strong B field normal to the 2DES layer, the Coulomb interaction gives rise to the fractional quantum Hall effect [7] and enhanced Zeeman splitting [8] observed in DC transport experiments. For dynamic excitations, assuming a parabolic band without the spin-orbit coupling, the single-particle transitions described above are replaced by collective excitations with the dispersion relations [9]:

$$E_{\Delta N, \Delta S}(q) = \Delta N \omega_c - \Delta S \omega_z + \Delta E_{\Delta N, \Delta S}(q), \qquad (4)$$

where the excited states are labeled by ΔN, ΔS, and wave vector q. For the $q = 0$ excitations using the THz or the GHz radiation, many-body corrections for CR and ESR are given by $\Delta E_{1,0}(0) = \Delta E_{0,-1}(0) = 0$, according to Kohn's theorem [10] and Larmor's theorem [9], which apply to translationally and spin rotationally invariant systems, respectively. In these cases, CR and ESR can be viewed in the many-body picture as the magnetoplasmon and spin wave at $q = 0$, respectively. On the contrary, no simple symmetry argument exists for the many-body correction $\Delta E_{1,-1}(0)$ to the CBR at $q = 0$. The shifted CBR is therefore labeled as the spin-flip excitation [11]. Via a spectroscopic experiment one can measure not only directly band splitting due to the spin-orbit coupling, but also get an easy access to many-body effects by analysing the oscillator strengths [12]. The main challenge is to improve the spectroscopic sensitivity.

Due to the bolometric effect [13] dipole excitation of electrons effectively heats the 2DEG and change its resistance. In the case of a weak FIR illumination and by applying a small bias current so that non-resonant heating of both the 2DEG and lattice can be neglected, the amplitude of the photo-voltage is determined by the steady state of the hot electron gas formed under the condition that its energy loss rate is equal to its power absorption. It is expressed as [13]

$$|\Delta V_{xx}| = I \cdot |\Delta R_{xx}| = I \cdot \left| \frac{\partial R_{xx}}{\partial T} \right| \cdot \frac{P(\omega) \cdot \tau_e}{C_e}, \qquad (5)$$

where I is the bias current, ΔR_{xx} the photo-resistance, $P(\omega)$ the absorbed power of the FIR radiation, and τ_e is the energy relaxation time of the hot electron gas. In the low-temperature limit $k_B T \ll E_F$ the 2DEG's heat capacity $C_e = \pi^2 k_B^2 T D(E_F, B)/3$ [14] is directly proportional to the temperature T and the density of states $D(E_F, B)$ at the Fermi energy E_F. Electron heating is therefore most effective in the vicinity of integer filling factors where C_e is small [15].

Over the decades, FIR spectroscopy, done mostly in simple transmission experiments has been used extensively to study charge excitations in electronic systems with reduced dimensionality [16]. The success of the technique lies in the fact that the characteristic energies of such charge excitations are typically between 1 to 100 meV, corresponding to photo frequencies in the FIR with $\bar{v} = \omega/2\pi c$ ranging from 10 to 1000 cm^{-1}. Strong spin-orbit coupling raises spin excitations from microwave up to FIR regime. These excitations are much weak compared to charge excitations. Therefore, far-infrared photoconductivity (FIR-PC) spectroscopy based on bolometric effect that has a much higher sensitivity than transmission spectroscopy is a powerful tool to investigate spin effects in such systems.

In this paper, we summarize our recent experiments, mainly performed on InGaAs/InAlAs and GaAs/AlGaAs heterostructures, to illustrate how FIR-PC can help us to increase our understanding on spin effects in electronic systems with reduced dimensionality. Among others, the spin-orbit coupling, which was discovered about 80 years ago by atomic spectroscopy and gave birth to the very concept of *spin*, has been found rather difficult to be spectrally measured for the 2DEGs. We solved the problem by measuring the CBR using FIR-PC spectroscopy [17]. The high sensitivity of the technique has the potential to study spin excitations in nanostructured semiconductors. As the first step, we investigated the role of the elementary excitations of an antidot array on its photoconductivity [18]. In contrast to the transmission spectroscopy, FIR-PC spectroscopy probes also the relaxation process of the photo-generated carriers, based on which we found a novel bolometric spin effect in a 2DEG around Landau level filling $v = 1$. It gives rise to a substantial resistance change that depends on the spin polarisation of the density of states of the 2DEG at the Fermi level [19].

EXPERIMENTAL DETAILS

Our experiments were performed by applying a DC current to a 2DEG Hall bar and measuring the changes of the longitudinal voltage drop caused by the FIR radiation. Broadband radiation from a mercury lamp was used as the FIR source. Two kinds of experiments (Fig. 1) were performed using different ways to modulate the FIR radiation. In the photo-conductivity

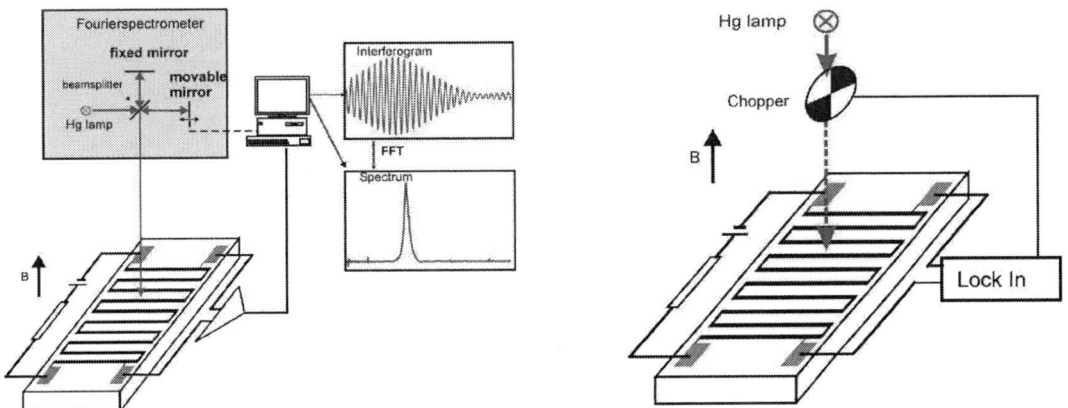

Figure 1. Experimental set up for FIR-PC spectroscopy with a Fourier transform spectrometer (left) and photo-modulated transport with lock-in amplify (right) on semiconductor samples with meandering long Hall bar.

spectroscopy experiment, the magnetic field B was fixed. Frequency domain spectra were obtained by using a Fourier-transform spectrometer with the sample itself as the detector. Here, the FIR radiation was modulated by the Michelson interferometer of the spectrometer, the corresponding change in the voltage drop of the sample was AC coupled to a broadband preamplifer and recorded as an interferogram, which can be Fourier-transformed to get the photo-conductivity spectrum. A 9 V battery and a load resistance R connected in series with the sample was used as the current source to avoid 50 Hz noise. In the photo-modulated transport experiment, a standard current source was used. The broadband FIR radiation was chopped at 16 Hz, and the photo-resistance was detected by a lock-in amplifer while sweeping the B field. In addition, we performed conventional transmission spectroscopy experiment using a Si bolometer as the detector, which allows us to monitor the phase correction factor sometimes needed for Fourier-transforming the photo-conductivity interferogram. The sample was mounted in a He cryostat with a superconducting solenoid in Faraday geometry.

Our GaAs sample is a modulation-doped $Al_{0.3}Ga_{0.7}As$/GaAs single heterojunction with a layer sequence: 1 µm buffer, 50 nm spacer, 57 nm $Al_{0.3}Ga_{0.7}As$ layer doped with Si to 2×10^{18} cm^{-3}, and 5 nm GaAs cap. The carrier density N_s and mobility μ at 4.2 K were determined by Shubnikov-de Haas (SdH) measurements to be 1.22×10^{11} cm^{-2} and 620 000 cm^2/Vs, respectively. The low density allows us to study the spin effects around the filling factor $v = 1$ where the 2DEG is spin polarized in the external B field.

For investigation of the spin-orbit coupling effect, we use an inverted-doped InAs step quantum well with 40-nm $In_{0.75}Al_{0.25}As$ cap layer. The step quantum well shown in Fig.4(a) is composed of 13.5-nm $In_{0.75}Ga_{0.25}As$, an inserted 4-nm InAs channel, and a 2.5-nm-thick $In_{0.75}Ga_{0.25}As$ layer. Underneath the quantum well is a 5-nm spacer layer of $In_{0.75}Al_{0.25}As$ on top of a 7-nm-wide Si-doped $In_{0.75}Al_{0.25}As$ layer. The sample is grown by molecular-beam epitaxy on a buffering multilayer accommodating the lattice mismatch to the semi-insulating GaAs substrate. A self-consistent Schrödinger Poisson calculation shows that the 2DES is about 55 nm below the surface, mainly confined in the narrow InAs channel. In all our samples, an extremely

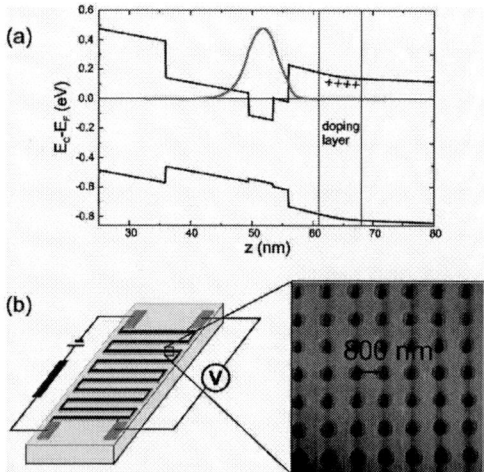

Figure 2. (a). Band structure for the InAs samples calculated by solving the Schrödinger and Poisson equations self-consistently. The 2DEG is formed about 55 nm below the surface, mainly confined in the narrow InAs channel. (b) Schematic bias circuit and sample structure showing the long Hall bar with ohmic contacts. The zoom-in part is an atomic force micrograph of the antidot array (from Ref. [18]).

long 2DEG Hall bar with a channel width of $W = 40$ μm and a total length L of about 10 cm was defined by chemical wet etching. The 2DEG channel runs meandering in a square of 4×4 mm^2. The extremely large L/W ratio enhances the sensitivity of our measurement. Ohmic contacts were made by depositing an AuGeNi alloy followed by annealing. For the antidot sample shown in Fig. 2(b), the long Hall bar was further patterned with the antidot array with a period of $a = 800$ nm. The holes with a geometric diameter of about 200 nm were defined by holography and chemical wet etching. The carrier density N_s and mobility μ at 2.2 K were determined by Shubnikov–de Haas measurement to be 6.66×10^{11} cm^{-2} and 150 000 cm^2/Vs, respectively, which reduce to 6.01×10^{11} cm^{-2} and 62 000 cm^2/Vs after fabricating the antidot array.

DISCUSSION

Combined resonance: measuring the spin-orbit coupling by FIR-PC spectroscopy

The combined resonance with the frequency described in Eq. (3) provides a direct measure of the spin-orbit coupling parameter. It is difficulty to be detected by FIR transmission spectroscopy due to its small matrix element. The advantage of FIR-PC spectroscopy is shown in Fig. 3 in which we compare spectra measured [17] on an InAs 2DEG by FIR-PC (solid lines) and transmission (dotted lines) experiment at two B fields. The transmission spectroscopy probes the high-frequency conductivity of the 2DEG so that the resonance strength is limited by the transition matrix element and the electron density [16]. On the contrary, photoconductivity of the 2DEG is caused by the bolometric effect where photoexcitation of electrons effectively heats the 2DEG, which changes its resistance. Its resonance intensity depends on the absorbed power of the radiation, the energy relaxation time of the photoexcited non-equilibrium electrons and the heat capacity of the 2DEG. Therefore, its sensitivity can be enhanced by carefully choosing the

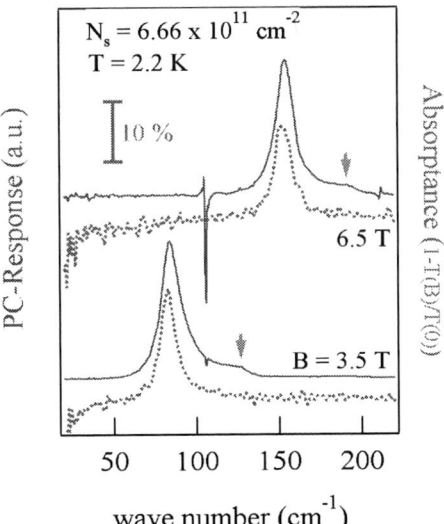

Figure 3. FIR photoconductivity spectra (solid lines) measured at two magnetic fields in comparison with conventional transmission spectra (dotted lines) under the same experimental conditions using a Si bolometer. In addition to the CR, arrows indicate the weak CBR which are only observable using the high sensitive photoconductivity technique (from Ref. [17]).

experimental condition and the sample design. The weak CBR appears at the high-energy side of the dominant CR peak, whose intensity is only about 0.8% of that of the CR, is only resolved in the highly sensitive photoconductivity spectrum.

Results of a systematic study of the B-field dispersion of both CR and CBR are summarized in Fig. 4(a). Also shown is the magnetoresistance R_{xx} measured without the FIR radiation with the same excitation current of 4.5 μA. The open circles determined from the CR are fit (dashed line) with Eq. (1) using m* = 0.039 m_e. The solid circles for the weak CBR are fit (dashed curve) using Eq. (3) with two fitting parameters Δ_R =38 cm^{-1} and g = -8.7. Observing the dipole-excited CBR in the Faraday configuration requires the spin-orbit interaction [4], in accordance with the obtained zero-field spin splitting $2\Delta_R$ = 76 cm^{-1}, which gives α = 2.38 ×10^{-11} eVm that is comparable to that measured early in similar samples using transport technique [5]. Using these parameters, we calculate the Landau levels and plot them in Fig. 4(b) together with the dotted lines showing the Fermi level. The arrows illustrate the CR and CBR, respectively. Note that the crossover of the Landau levels with opposite spin at small B-field regime is the characteristic of the spin-orbit coupling. It could give rise to a beating pattern in the R_{xx} curve but was not observed in our sample. This phenomenon has been pointed out in a previous study and is now the major mystery questioning the strength of the spin-orbit coupling [20]. Our results shed light on the mystery indicating clearly that missing a beating pattern does not simply mean that the spin-orbit coupling vanishes.

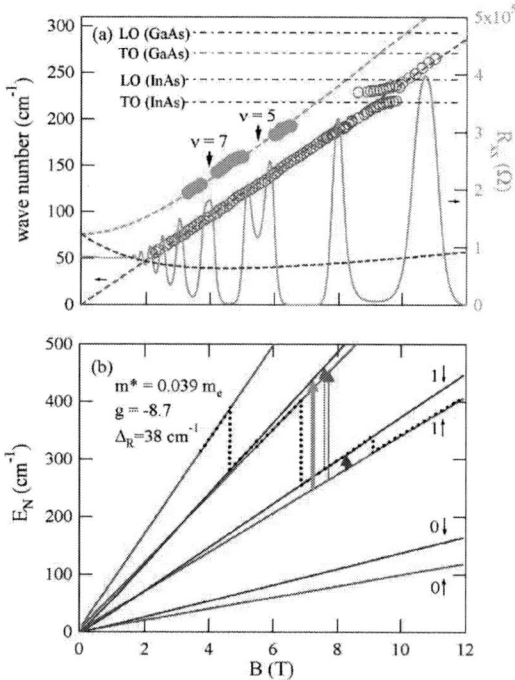

Figure 4. (a) Resonance dispersions determined from the photo-conductivity spectra and magnetoresistance R_{xx} measured without the FIR radiation. The dashed line and curve are fits for the CR and CBR using Eq. (1) and (3), respectively. Dash-dotted lines indicate the optical-phonon energies of InAs and GaAs. (b) Landau levels calculated using the band parameters obtained from the fits in (a). Dotted lines indicate the Fermi energy. The arrows illustrate the transitions for CR, ESR and CBR.

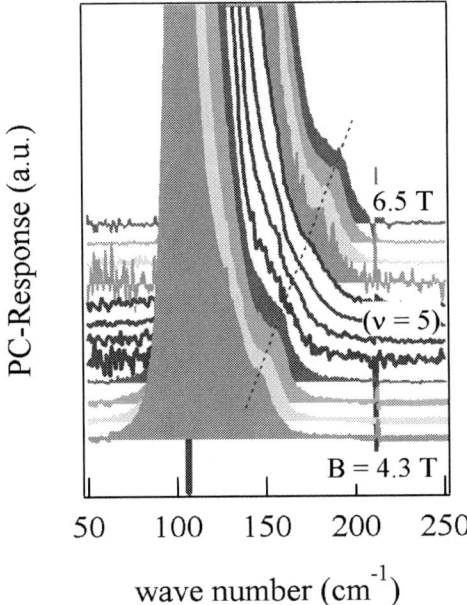

Figure 5. Normalized photoconductivity spectra measured between 4.3 and 6.5 T displaying damping of the CBR around $v = 5$, indicating that it should be better understood as a collective spin-flip excitation. The dotted line is guide to eyes. (from Ref. [17]).

The many-body character of the CBR can be found from the study of its normalized resonant strength. We plot in Fig. 5 the blown-up spectra measured between 4.3 and 6.5 T around $v = 5$. All spectra are normalized using their CR so that we can directly compare the intensity of the CBR without worrying about the change of the photoconductivity sensitivities at different B fields. We find clearly that the CBR disappears around $v = 5$. The same behavior is observed around $v = 7$. It provides us with the strongest evidence that the CBR we observe should be better understood as the collective spin-flip excitation, since theory [21] has long predicted that at odd filling factors where the ground state of the 2DEG is spin polarized, the collective spin-flip excitation decays into a magnetoplasmon and a spin wave that conserve spin, momentum, and energy. Strictly speaking, to accurately determine the g factor and spin-orbit parameter α, the electron-electron and spin-orbit coupling should be both taken into account. However, such a unified picture that is essential to understand the spin effects in the 2DES of InAs is not yet available.

Elemental excitations in an antidot array: towards FIR-PC of nanostructures

As a part of our efforts to study the charge and spin excitations in nanostructured semiconductors, we have performed FIR-PC spectroscopy experiment on an antidot array embedded in the meandering long Hall bar on InAs sample [18] as shown in Fig. 2(b). Figure 6(a) shows typical FIR-PC spectra measured at different B fields. At the low magnetic field of $B = 3.2$ T, two resonances are clearly observed. These are the characteristic collective excitations of an antidot array subjected to a perpendicular magnetic field B [22]. By increasing the B field to

6.4 T, the lower-energy edge-magnetoplasmon mode ω_{EMP} has a slight red shift and gets weaker, while the higher-energy ω^+ mode shows a significant blue shift and dominates the spectrum. Two additional weak resonances are observed: one appears as a shoulder (thick arrow) of the dominant resonance and the other lies at about 285 cm^{-1} (thin arrow). By further increasing the B field to 9.6 T, the ω^+ mode splits into multi-peaks, while the weak structure at 285 cm^{-1} gains resonance strength, both are effects related with optical phonons in the multi-layer semiconductor structure. The edge-magnetoplasmon mode disappears at large B fields.

Fig. 6(b) shows the B-field dispersion of the resonances together with the magnetoresistance R_{xx} measured without FIR radiation. The ω^+-mode at large B field approaches the CR frequency. Both the ω^+ and ω_{EMP} can be nicely fit (solid curve) using a theory with modified-dipole and effective-medium approximations [23]. Within the fitting accuracy, the fit effective mass value with $m^* = 0.039\ m_e$ is equal to that directly measured from the CR on the unpatterned sample which we have described before. By comparing the spectra with that obtained on the unpatterned sample, we further identify the weak resonance marked by the thick arrow in Fig. 6(a) as the collective spin-flip excitation. With the antidot lattice, the collective spin-flip excitation gets broader and appears as a shoulder of the ω^+ mode. To our knowledge, spin-flip excitation in an antidot array has neither been studied experimentally nor investigated theoretically. Here we have assumed that at large B fields, the spin-flip excitation in the antidot array approaches that in an unpatterned 2DEG, just like the ω^+ mode approaches the CR. For comparison, in Fig. 6(b) we plot the calculated dispersion for the 2DEG spin-flip excitation neglecting both the many-body correction and the antidot potential, using $\alpha = 2.38 \times 10^{-11}$ eVm and $g = -8.7$ determined for the unpatterened sample. Within the B-field range of 6–7 T, where we can observe the spin-flip excitation, the influence of the antidot potential on its resonance frequency is found small. These results show the potential of applying FIR-PC to study both charge and spin excitations in nanostructured semiconductors.

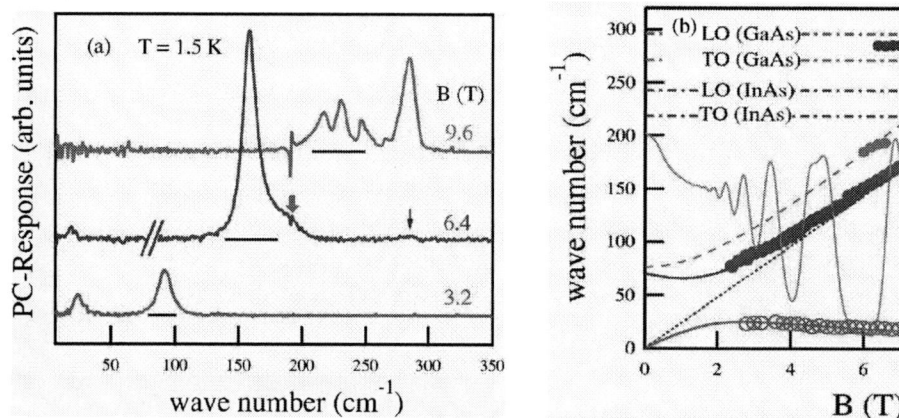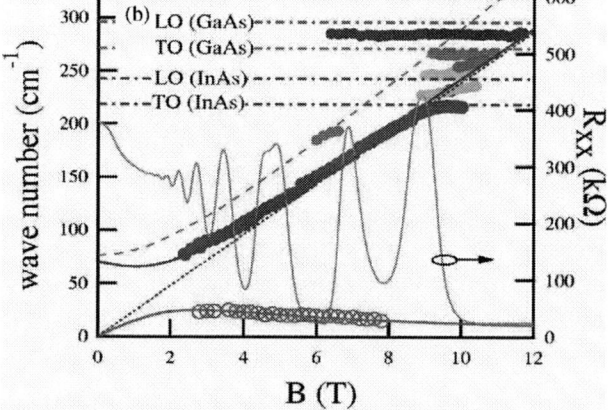

Figure 6. (a) FIR photoconductivity spectra measured at $T = 1.5$ K at different B fields. Thick arrow indicate the spin-flip excitation. Spectra are vertically offset for clarity. (b) B-field dispersions for resonances displayed in (a) and magnetoresistance R_{xx} measured without FIR radiation. The solid curves are theoretical fits for ω^+ (solid circles) and ω_{EMP} (open circles) modes. The dotted line and dashed curve are calculated for CR and CBR, respectively, using $m^* = 0.039\ m_e$, $\alpha = 2.38 \times 10^{-11}$ eVm and $g = -8.7$. Dash-dotted lines indicate the optical phonon energies of InAs and GaAs. (from Ref. [18]).

Bolometric spin effect: influence of spin relaxation on FIR-PC

In addition to the high sensitivity advantage, FIR-PC also provides the access to the spin-relaxation process that can not be probed by transmission experiments. Figure 7 shows the FIR-PC spectra measured on a GaAs 2DEG at 2.1 K with a bias current of 0.9 μA under different B fields [19]. At those B fields with ν around 1 and 2, clear resonance spectra were observed, indicating large photo-resistance of the 2DEG. According to eq. (5), it means that the bolometric effect is most effective in the vicinity of integer filling factors where C_e is small. In the inset, the spectrum measured at ν = 1 by FIR-PC spectroscopy is scaled down to compare with the CR measured by transmission spectroscopy, which shows nearly identical Lorentzian line shape. It allows us to identify the resonance as resistively detected CR of the 2DEG.

Within the narrow band of the CR frequency around each integer filling factor, $P(\omega)$ in eq. (5) can be regarded as frequency independent. If we assume that τ_e does not strongly depend on the filling factor, the B-field dependence of the photo-resistance described by eq. (5) will be determined by $D(E_F, B)$ and $\partial R_{xx}/\partial T$. By taking into account the disorder-induced broadening of the Landau levels, $D(E_F, B)$ is known to be symmetric against each integer filling factor [14], while $\partial R_{xx}/\partial T$ can be directly measured. In fig. 8(a) we plot the magnetoresistance R_{xx}, measured without FIR radiation at $I = 0.45\ \mu A$, together with the value of $\partial R_{xx}/\partial T$ obtained by measuring $[R_{xx}(2.3K)-R_{xx}(2.05K)]/0.25K$. We find that $\partial R_{xx}/\partial T$ displays a double-peak around both ν = 1 and 2, which explains well the filling factor dependence of the resonance amplitude measured around ν = 2 shown in Fig. 7. In contrast, the filling factor dependence of the photo-resistance $|\Delta R_{xx}|$ around ν = 1 is completely different. To avoid any uncertainty in determining $|\Delta R_{xx}|$ due to the Fourier transform of the interferograms, we measured it directly in the photo-modulated transport experiment. In fig. 8 (b) we plot $|\Delta R_{xx}|$ measured at bias currents of $I < 1\ \mu A$ without severely heating the 2DEG. The double-peak behavior around ν=2 is now more clearly presented. Apparently, $|\Delta R_{xx}|$ around ν=1 shows a very different behavior. The peak at the low B field side with ν=1⁺ is much larger (more than a factor of 20) than that at ν =1⁻.

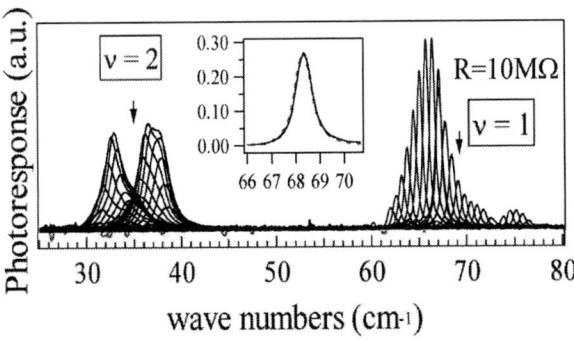

Figure 7. FIR-PC spectra measured on GaAs 2DEG at 2.1 K with $I = 0.9\ \mu A$ under different B fields. The inset shows the comparisons of CR measured in FIR-PC (solid lines) and transmission (dashed lines) spectroscopy at $B = 5$ T, corresponding to ν = 1. (from Ref. [19]).

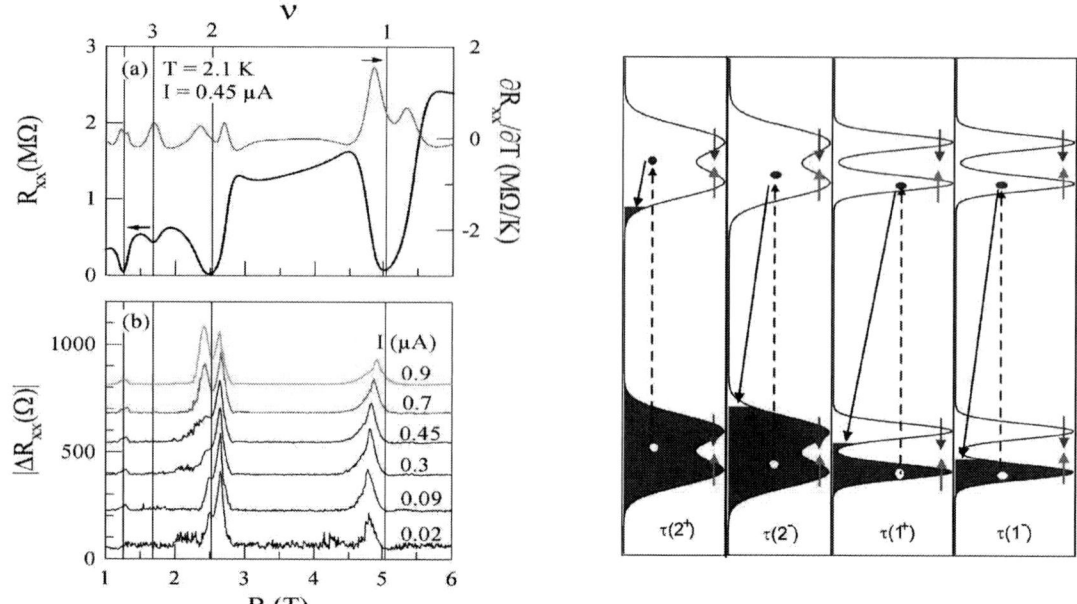

Figure 8. (left) (a) Magnetoresistance R_{xx} and $\partial R_{xx}/\partial T$ measured with $I = 0.45 \,\mu A$. (b) Photo-resistance $|\Delta R_{xx}|$ measured with different bias currents. The curves are vertically shifted for clarity. The dotted vertical lines indicate the filling factors. (from Ref. [19]).

Figure 9. (right) Schematic diagrams of the CR excitation and relaxation processes at Landau-level filling $\nu = 2^+, 2^-, 1^+$ and 1^-. Please note the spin-flip process at $\nu = 1^+$. (from Ref. [19]).

To understand such a striking filling factor dependence around $\nu = 1$, we need to examine the picture of the inter-Landau-level relaxation of the CR-excited electrons. There are two possible relaxation processes. One is the radiative recombination with the relaxation time determined by the dipole matrix element for CR which does not change much around integer filling factors. The other one is non-radiative relaxation where the CR-excited electron loses energy to the electron gas and the lattice via electron-electron and electron-phonon interaction, respectively, after which it is relaxed to an empty state at the Fermi level. Equation (5) applies for conventional bolometric effect, where the energy equilibration among electrons is faster than the competing direct energy loss to the lattice [13], so that the steady state of the hot electron gas is defined by a temperature change of $\Delta T = P(\omega) \cdot \tau_e / C_e$ depending on the energy relaxation time τ_e of the hot electron gas.

The significant difference between the bolometric effect around $\nu = 2$ and 1 is caused by their different spin nature. In fig. 9 we illustrated the CR excitation and its subsequent non-radiative relaxation process with the dashed and solid arrows, respectively. Because dipole transitions conserve the spin quantum number, CR-excited electrons around $\nu = 1$ are spin polarised. Their relaxation depends on the spin-dependent density of states at the Fermi level, which changes around $\nu = 1$. At $\nu = 1^+$, inter-Landau-level relaxation of the excited electrons must be accompanied by a spin-flip scattering, while for $\nu = 1^-$ it must not. If the spin-relaxation time is very short, so that the inter-Landau-level relaxation is faster than the time constant τ_e the bolometric response is just determined by the steady state of the hot electron gas with the temperature change ΔT, for which τ_e is the relevant time constant. If the spin-relaxation time is so long that the energy equilibration among electrons cannot be separated from the direct energy loss process to the lattice, it is not possible to define the steady state by an electron temperature using the time constant τ_e. In contrast to the conventional bolometric effect, $|\Delta R_{xx}|$ no longer simply follows $\partial R_{xx}/\partial T$ but depends on the details of the energy distribution of the 2DEG at the steady state. Indeed, it has been found that the spin-relaxation time of the optically excited electrons of a 2DEG can be extremely long with an electron spin temperature being driven above the lattice temperature [24]. We therefore attribute the large value of $|\Delta R_{xx}(1^+)|/|\Delta R_{xx}(1^-)|$ to a distinct bolometric spin effect, where the conventional bolometric effect is remarkably changed if both the optically excited electrons and the density of states of the 2DEG are spin polarised. Around $\nu = 2$, CR-excited electrons are not spin polarised, therefore $|\Delta R_{xx}(2^+)|/|\Delta R_{xx}(2^-)|$ is of the order of unity, as predicted by the conventional bolometric effect. We note that the bolometric spin effect could provide us with a unique tool to quantitatively investigate the complicated but interesting nature of spin-relaxation processes of the 2DEG via the resistance measurement.

CONCLUSIONS

We have illustrated a few examples to demonstrate how spin effects in electronic systems with reduced dimensionality can be investigated by FIR-PC experiments. In these systems, charge effects have been intensively studied by FIR transmission spectroscopy over the decades, which have significantly increased our understanding of quantum physics. We expect that FIR-PC will continue to give us a detailed insight into the spin physics in these systems. With the advantages such as extremely high sensitivity as well as the access to the spin-relaxation process, many intriguing spin effects seems possible to be investigated by this technique, such as the dynamic coupling between charge and spin excitations, interplay between spin-orbit coupling and electron-electron interaction, as well as the impact of long-wave length spin excitations on the long-range magnetic order. Such studies might bring us to new directions in the exploding field of semiconductor spintronics [25].

ACKNOWLEDGMENTS

This invited contribution has been based on collaborative studies which I unfortunately unable to present at the MRS 2004 spring meeting in San Francisco due to the "meandering long" visa process. I want to thank C. Zehnder, A. Wirthmann, K. Bittkau, Ch. Menk, Ch. Heyn, D. Heitmann and Y.S. Gui for their valuable contributions. I also wish to acknowledge financial supports by the EU 6th-Framework Programme through project BMR, the NEDO international spintronics project, the DFG through SFB 508 and BMBF through project 01BM905.

REFERENCES

1. G. Abstreiter, P. Kneschaurek, J.P. Kotthaus, and J.F. Koch, Phys. Rev. Lett. **32**, 104 (1974); S.J. Allen, Jr., D.C. Tsui, and J.V. Dalton, Phys. Rev. Lett. **32**, 107 (1974).

2. D. Stein, K.v. Klitzing, and G. Weimann, Phys. Rev. Lett. **51**, 130 (1983).
3. B. Das, S. Datta, and R. Reifenberger, Phys. Rev. B **41**, 8278 (1990).
4. B.D. McCombe, S.G. Bishop, and R. Kaplan, Phys. Rev. Lett. **18**, 748 (1967); B.D. McCombe, Phys. Rev. **181**, 1206 (1969).
5. J. Nitta *et al.*, Phys. Rev. Lett. **78**, 1335 (1997); G. Engels *et al.*, Phys. Rev. B **55**, 1958 (1997); C.-M. Hu *et al.*, *ibid.* **60**, 7736 (1999); T. Matsuyama *et al.*, *ibid.* **61**, 15 588 (2000); D. Grundler, Phys. Rev. Lett. **84**, 6074 (2000).
6. S. Datta and B. Das, Appl. Phys. Lett. **56**, 665 (1990).
7. T. Chakraborty and P. Pietiläinen, *The Quantum Hall Effects: Fractional and Integral,* (Springer, Berlin, MA, 1995).
8. T. Ando, A.B. Fowler, and F. Stern, Rev. Mod. Phys. **54**, 437 (1982).
9. C. Kallin and B.I. Halperin, Phys. Rev. B **30**, 5655 (1984); A. H. MacDonald and C. Kallin, Phys. Rev. B 40, 5795 (1989).
10. W. Kohn, Phys. Rev. **123**, 1242 (1961).
11. A. Pinczuk, B.S. Dennis, D. Heiman, C. Kallin, L. Brey, C. Tejedor, S. Schmitt-Rink, L.N. Pfeiffer, and K.W. West, Phys. Rev. Lett. **68**, 3623 (1992).
12. C.-M. Hu, E. Batke, K. Köhler, and P. Ganser, Phys. Rev. Lett. **75**, 918 (1995); Phys. Rev. Lett. **76**, 1904 (1996).
13. F. Neppl, J.P. Kotthaus, and J.F. Koch, Phys. Rev. B **19**, 5240 (1979).
14. J.K. Wang, D.C. Tsui, M. Santos, and M. Shayegan, Phys. Rev. B **45**, 4384(1992).
15. K. Hirakawa, K. Yamanaka, M. Endo, M. Saeki, and S. Komiyama, Phys. Rev. B **63**, 085320 (2001).
16. J.P. Kotthaus, "Infrared excitations in electronic systems with reduced dimensionality", *NATO ASI Series, Interface, quantum wells, and superlattices,* ed. C.R. Leavens and R. Taylor (Plenum Press, New York and London, 1988) pp. 95-126.
17. C. -M. Hu, C. Zehnder, Ch. Heyn, D. Heitmann, Phys. Rev. B, **67**, 201302 (R) (2003).
18. K. Bittkau, Ch. Menk, Ch. Heyn, D. Heitmann, and C. -M. Hu, Phys. Rev. B, **68** 195303 (2003).
19. C. Zehnder, A. Wirthmann, Ch. Heyn, D. Heitmann and C. -M. Hu, Europhys. Lett., **63**, 576 (2003).
20. S. Brosig, K. Ensslin, R.J. Warburton, C. Nguyen, B. Brar, M. Thomas, and H. Kroemer, Phys. Rev. B **60**,13 989(1999).
21. J.P. Longo and C. Kallin, Phys. Rev. B **47**, 4429 (1993).
22. K. Kern, D. Heitmann, P. Grambow, Y.H. Zhang, and K. Ploog, Phys. Rev. Lett. **66**, 1618 (1991); A. Lorke, J.P. Kotthaus, and K. Ploog, Superlattices Microstruct. **9**, 103 (1991); Y. Zhao, D.C. Tsui, M. Santos, M. Shayegan, R.A. Ghanbari, D.A. Antoniadis, and H.I. Smith, Appl. Phys. Lett. **60**, 1510 (1992).
23. S.A. Mikhailov and V.A. Volkov, Phys. Rev. B **52**, 17260 (1995); S.A. Mikhailov, *ibid.* **54**, 14293 (1996).
24. N.N. Kuzma, P. Khandelwal, S.E. Barrett, L.N. Pfeiffer, and K.W. West, Science **281**, 686 (1998); J.M. Kikkawa and D.D. Awschalom, Nature **397**, 139 (1999).
25. I. Žutić, J. Fabian, and S. Das Sarma, Rev. Mod. Phys. **76**, 323 (2004).

Mat. Res. Soc. Symp. Proc. Vol. 825E © 2004 Materials Research Society

Spin-orbit coupling and magnetic spin states in cylindrical quantum dots

C. F. Destefani,[1,2] Sergio E. Ulloa,[1] and G. E. Marques[2]

[1] *Department of Physics and Astronomy, Ohio University, Athens, Ohio 45701-2979*
[2] *Departamento de Física, Universidade Federal de São Carlos, 13565-905, São Carlos, São Paulo, Brazil*

We make a detailed analysis of each possible spin-orbit coupling of zincblende narrow-gap cylindrical quantum dots built in a two-dimensional electron gas. These couplings are related to both bulk (Dresselhaus) and structure (Rashba) inversion asymmetries. We study the competition between electron-electron and spin-orbit interactions on electronic properties of 2-electron quantum dots.

PACS numbers: 71.70.Ej, 73.21.La, 78.30.Fs
Keywords: spin-orbit coupling, Rashba effect, quantum dots

The creation and manipulation of spin populations in semiconductors has received great attention since the Datta-Das proposal of a spin field-effect transistor,[1] based on Rashba spin-orbit coupling of electrons in a bidimensional electron gas,[2] and the possibility for quantum computation devices using quantum dots (**QDs**).[3] Thus, it is important that every spin-orbit (**SO**) effect be clearly understood for a full control of spin-flip mechanisms in nanostructures.

There are two main **SO** contributions in zincblende materials. In addition to the structure inversion asymmetry (**SIA**) caused by the 2D confinement (the Rashba **SO**), there is a bulk inversion asymmetry (**BIA**) term in those structures (the Dresselhaus **SO**).[4] An yet additional lateral confinement defining a dot introduces another **SIA** term with important consequences, as we will see in detail. Although the relative importance of these two effects depends on the material and on sample design (via interfacial fields), only recently have authors begun to consider the behavior of spins under the influence of all effects.

The goal of this work is to show how important different types of **SO** couplings are on the spectra of parabolic **QDs** built with narrow-gap zincblende materials. We consider both Rashba and a diagonal **SIA**, as well as the all Dresselhaus **BIA** terms in the Hamiltonian, in order to study features of the spectrum as function of magnetic field, dot size, and electron-electron interaction.

Consider a heterojunction or quantum well confinement potential $V(z)$ such that only the lowest z-subband is occupied. The Hamiltonian for a cylindrical **QD**, in the absence of **SO** interactions, is given by $H_0 = (\hbar^2/2m)\mathbf{k}^2 + V(\rho) + g\mu_B \mathbf{B} \cdot \boldsymbol{\sigma}/2$, where $\mathbf{k} = -i\boldsymbol{\nabla} + e\mathbf{A}/(\hbar c)$, $\mathbf{A} = B\rho(-\sin\theta, \cos\theta, 0)/2$ describes a magnetic field $\mathbf{B} = B\mathbf{z}$, m is the effective mass in the conduction band, g is the bulk g-factor, μ_B is Bohr's magneton, $V(\rho) = m\omega_0^2\rho^2/2$ is the lateral dot confinement, and $\boldsymbol{\sigma}$ is the Pauli spin vector. The analytical solution of H_0 yields the Fock-Darwin (**FD**) spectrum, $E_{nl\sigma_z} = (2n + |l| + 1)\hbar\Omega + l\hbar\omega_C/2 + g\mu_B B\sigma_z/2$, with effective (cyclotron) frequency $\Omega = \sqrt{\omega_0^2 + \omega_C^2/4}$ ($\omega_C = eB/(mc)$). The **FD** states are given in terms of Laguerre polynomials.[5] The lateral, magnetic and effective lengths are $l_0 = \sqrt{\hbar/(m\omega_0)}$, $l_B = \sqrt{\hbar/(m\omega_C)}$ and $\lambda = \sqrt{\hbar/(m\Omega)}$, respectively.

The **SIA** terms[2] for the full confining potential, $V(\mathbf{r}) = V(\rho) + V(z)$, and coupling parameter α are decomposed as $H_{SIA} = H_R + H_{SIA}^D + H_K$. These three forms are: i) $H_K = i\alpha(\hbar\omega_0/l_0^2)\lambda x[\sigma_+ L_- - \sigma_- L_+]\langle k_z\rangle$ gives zero contribution when $\langle k_z\rangle = 0$ (pure state parity); ii) $H_{SIA}^D = \alpha\sigma_z(\hbar\omega_0/l_0^2)[L_Z + \lambda^2 x^2/2l_B^2]$ is the diagonal contribution due to the lateral confinement and $x = \rho/\lambda$ been a dimensionless radial coordinate, $L_Z = -i\partial/\partial\theta$ is the z-orbital angular momentum; iii) $H_R = -\alpha(dV/\lambda dz)[\sigma_+ L_- A_- + \sigma_- L_+ A_+]$ is the Rashba term for the perpendicular confinement dV/dz, $L_\pm = e^{\pm i\theta}$, $\sigma_\pm = (\sigma_X \pm i\sigma_Y)/2$ and $A_\pm = [\mp\partial/\partial x + L_Z/x + x\lambda^2/(2l_B^2)]$. In principle these terms can be tuned since H_{SIA}^D depends on the confining frequency ω_0 while H_R depends on the interfacial field dV/dz.

The **BIA** Hamiltonian[4] for zincblende materials, after averaging along the z-direction, is given by $H_{BIA} = \gamma\left[\sigma_x k_x k_y^2 - \sigma_y k_y k_x^2\right] + \gamma\langle k_z^2\rangle\left[\sigma_y k_y - \sigma_x k_x\right] + \gamma\sigma_z\langle k_z\rangle\left(k_x^2 - k_y^2\right)$, where γ is the Dresselhaus parameter. The first (second) term is cubic (linear) in the in-plane momentum. The last term will be zero for systems where $\langle k_z\rangle = 0$, while $\langle k_z^2\rangle \simeq (\pi/z_0)^2$, z_0 being the z-direction (perpendicular) confinement length. In cylindrical coordinates H_{BIA} can be written as $H_{BIA} = H_D^L + H_D^C$, where $H_D^L = -i(\gamma/\lambda)[\sigma_+ L_+ A_+ - \sigma_- L_- A_-]\langle k_z^2\rangle$ is the linear term and, after long algebra manipulation,[6] the cubic term becomes $H_D^C = i(\gamma/\lambda^3)[\sigma_- L_+^3 H_1 + \sigma_+ L_-^3 H_2 + \sigma_- L_- H_3 + \sigma_+ L_+ H_4]$, where $H_i = A_i + \frac{\lambda^2}{l_B^2}B_i + \frac{\lambda^4}{l_B^4}C_i + \frac{\lambda^6}{l_B^6}D_i$, with $i = 1, 2, 3, 4$. The long expressions for the sixteen functions A_i, B_i, C_i, D_i are given in Ref. [6].

Finally the electron-electron interaction $H_{ee} = e^2/[\varepsilon|\mathbf{r}_1 - \mathbf{r}_2|]$, with ε being the dielectric constant of the material, is expanded into Bessel functions $J_k(\xi)$ as $H_{ee} = (\hbar\Omega\lambda/a_B)\sum_{k=-\infty}^{\infty} e^{ik(\theta_1-\theta_2)}\int_0^\infty d\xi J_k(\xi x_1)J_k(\xi x_2)e^{-\xi z_0/\lambda}$, where $a_B = \varepsilon\hbar^2/(me^2)$ is the effective Bohr radius. The **FD** basis states must be properly antisymmetrized to describe unperturbed spin eigenstates.[6]

Summarizing, our total single-particle Hamiltonian is given by $H = H_0 + H_{SIA}^D + H_R + H_D^L + H_D^C$. For the two-particle case, we study the states and spectrum of $H + H_{ee}$. Parameters for InSb are in Ref. [7].

37

G4.6.2

We present results by analyzing the role of each **SO** term in the Hamiltonian. We take into account all states in the **FD** basis having $n \leq 4$ and $|l| \leq 9$ in our numerical diagonalization, which is equivalent to the first ten energy shells at zero field and embodies a total of 110 basis states. The sequence of **FD** states of H_0 starts at zero B-field with $\{n, l, \sigma_Z\} \equiv \{0, 0, \pm 1\}$, followed by the degenerate set $\{0, -1, \pm 1\}$ and $\{0, 1, \pm 1\}$. The next energy shell is composed by $\{0, -2, \pm 1\}$, $\{1, 0, \pm 1\}$, and $\{0, 2, \pm 1\}$.[5] Spin and orbital degeneracies are broken by B and the states with negative l and positive σ_Z acquire lower energies because of the negative g-factor of InSb. The lowest **FD** crossing occurs between states $\{0, 0, -1\}$ and $\{0, -1, +1\}$, at a critical field $B_C^0 = \widetilde{m} \hbar \omega_0 / [\mu_B \sqrt{\widetilde{m}} |g| (\widetilde{m} |g| + 2)]$. The moderate value of B_C^0 in InSb ($\simeq 2.6$ T for $\hbar \omega_0 = 15$ meV) is a direct consequence of its large $|g|$ factor.[7] For GaAs ($|g| = 0.44$, $\widetilde{m} = m/m_0 = 0.067$), this level crossing would appear only at $B_C^0(GaAs) \simeq 9.5$ T for a much smaller confinement, $\hbar \omega_0 = 2$ meV, corresponding to a regime where Landau levels are well defined.

Any figure showing spectrum has the structure: Panel A shows **QD** spectrum for the full **FD** basis (110 states); Panel B shows a zoom on the three lowest shells, plus inset with another zoom on the 4 levels of the second shell; Panel C and D show, respectively, the B-evolution of spin σ_Z and orbital l angular momenta, for the full **FD** basis, while their insets take into account only the lowest 7 **QD** levels.

Figure 1 shows the simultaneous addition of both **SIA** terms, H_{SIA}^D and H_R, to H_0. The diagonal term H_{SIA}^D causes small splittings on the zero-field spectrum and readjust the sequence of states according to total angular momentum $j = l + \sigma_Z/2$. For example, the highest (lowest) zero-field energy level in the second shell has $j = 3/2$ ($j = 1/2$). Since H_{SIA}^D does not induce shift on the accidental degeneracy points of the **FD** spectrum at finite fields, the first level crossing occurs at a critical value $B_C \simeq B_C^0 \simeq 2.6$ T. Also, this term does not induce any level mixture on the **FD** states. The Rashba term H_R introduces a strong state mixture for any magnitude of α. This is evident for any pair of **FD** levels satisfying $\Delta l = -\Delta \sigma_Z/2 = \pm 1$ that show a crossing at given accidental degeneracy of the **FD** spectrum. The induced mixture converts this crossing at B_C^0 to an anticrossing (**AC**), at a shifted critical field $B_C \simeq 2.5$ T $\lesssim B_C^0$, with an energy minigap. Higher energy levels, also satisfying this selection rule, present **AC**s around the same value of B_C, and gives origin to the observed collapse in both σ_Z and l quantum numbers at $B \simeq 2.5$ T, shown in Panels C and D. The range of critical fields (between 2.1 and 2.6 T), and the size of the minigaps opened at those **AC** regions, are proportional to the magnitude of α. H_R also induces small splittings in the zero-field spectrum and slightly shifts the accidental degeneracy points at finite fields. After adding both **SIA** terms simultaneously (full spectrum in Panel A), one can see in the inset of Panel B that the ordering of states is the one determined by H_{SIA}^D. However, the energies of states $j = 3/2$ at 30 meV and $j = 1/2$ at a slightly smaller value, as well as the value of field ($B \gtrsim 0.1$ T) where the normal state ordering (the one existing in the absence of SO terms) for $g < 0$ material is restored, are determined by H_R. For increasing energy, the ordering for this second shell is $\{0, -1, +1\}$, $\{0, -1, -1\}$, $\{0, 1, +1\}$, $\{0, 1, -1\}$. The width of critical fields becomes wider, between 2.2 and 3.6 T, as seen in Panel C. Furthermore, Panel D shows that orbitals having $l < 0$ ($l > 0$) present **AC**s at fields smaller (larger) than the field $B_C \simeq 2.55$ T where occurred the first **AC** (see insets of Panels C and D). For a future comparison notice, in Panel B, that the first **AC** near 50 meV involving $n = 1$ states $\{0, 1, -1\}$ and $\{1, 0, +1\}$ occurs at that same B_C value. In general, **AC**s between states with any n value occur inside the same unique range of critical fields, as shown in Panels C and D. Finally, observe that both **SIA** terms can be reduced by decreasing ω_0 (H_{SIA}^D) or dV/dz (H_R), and that all even and negative l states ($l = -2, -4, -6, -8$) show anticrossings.

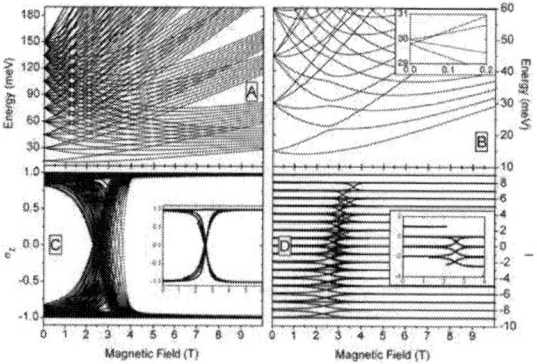

FIG. 1: Spectrum when H_{SIA}^D and H_R are added to H_0 (A and B). Critical field range for **AC**s is seen on C and D. The lowest one ($B_C \simeq 2.55$ T, insets) occurs near B_C^0. **AC**s involving $l < 0$ ($l > 0$) orbitals are shifted to lower (higher) fields (D).

38

Figure 2 shows the simultaneous addition of both **BIA** terms, H_D^C and H_D^L, to H_0. The cubic term H_D^C, under the present **QD** parameters, has small influence over the H_0 spectrum. Small state mixtures is induced at $B_0 \simeq 1$ and $\simeq 5$ T, both involving **ACs** satisfying $\Delta l = \mp 3$ and $\Delta \sigma_Z/2 = \pm 1$. The first one, at 1 T (5 T), occurs between states $\{0, 1, -1\}$ and $\{0, -2, +1\}$ ($\{0, 0, -1\}$ and $\{0, -3, +1\}$). This term also induces zero-field splittings on the **FD** spectrum and also a shift due to matrix elements for $\Delta l = \pm 1 = \Delta \sigma_Z/2$. However, such splittings and opened minigaps at the **ACs** regions are very small. Therefore, the simultaneous addition of both **BIA** terms (full spectrum in Panel A), where the linear term H_D^L is the most important, drastically changes the general features of the **FD** spectrum, inducing strong zero-field splittings and shifting its accidental degeneracies to higher fields. Yet, the respective matrix elements ($\Delta l = \pm 1 = \Delta \sigma_Z/2$) does not introduce **ACs** on the lowest energy levels. As seen in Panels C and D, the mixing induced by the linear term is so strong that the **QD** states are not anymore pure states even at zero field. Notice in Panel C that at $B = 0$ values of $|\sigma_Z| < 0.5$ are found for high energy states, while in its inset one finds $\sigma_Z \simeq 0.7$ for the ground state. As an example of level crossings displaced to higher fields, observe in Panel B that the first one has moved to $B_C \simeq 3.3$ T, there is only one crossing present in the second shell at about 0.45 T (see inset), and the second one occurs a higher field around 3.5 T. Thus, contrary to the observed for **SIA** case, the normal ordering of state is no longer restored. We will come back to this fact later. As a final note, in the same inset and at zero field, the highest (lowest) energy state has $j = 3/2$ with eigenvalue equal to energy of 30 meV of the pure H_0 ($j = 1/2$ at smaller energy near 27 meV). The influence of H_D^L on the spectrum changes with z_0.

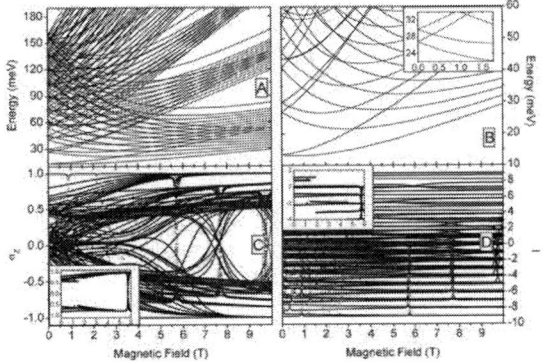

FIG. 2: Spectrum when H_D^C and H_D^L are added to H_0. The linear contribution dominates the cubic one. Strong mixing at low fields are due to H_D^L, while the **AC** around 6 T is due to H_D^C.

Figure 3 shows the one-particle **QD** spectrum for full H or when all **SO** terms are simultaneously taken into account. From the previous discussions, one may identify which **SO** mechanism is dominant in each of the main signatures present on the spectrum. An enormous state mixture, even at small magnetic fields (Panels C and D and their insets), as also splittings, position and ordering of states (Panels A, B and its inset) are dominated by H_D^L, although with contributions from both **SIA** terms. The small influence of H_D^C remains around 6 T. The lowest **ACs** are induced by the Rashba term H_R, although shifted to higher critical fields by the linear **BIA** term, H_D^L. Observe that the first **AC** has moved from 2.55 T (Fig. 1) to 3.3 T (Panel B and insets of Panels C and D), and the ranges of critical fields becomes wider. However, the new feature of the full H spectrum is the clear presence of more than one unique range of critical fields where **ACs** occur (compare with Fig. 1). Details on Panels A and C: i) The first family of **ACs**, near 3.3 T (related to states between 20 and 70 meV) involves only $n = 0$ levels. The first **ACs** being between $\{0, 0, -1\}$ and $\{0, -1, +1\}$, $\{0, -1, -1\}$ and $\{0, -2, +1\}$, $\{0, -2, -1\}$ and $\{0, -3, +1\}$, ... ; ii) A second family of **ACs** around 5 T (related to states between 70 and 120 meV) involves only $n = 1$ levels. The first **ACs** being between $\{0, 1, -1\}$ and $\{1, 0, +1\}$, $\{1, 0, -1\}$ and $\{1, -1, +1\}$, $\{1, -1, -1\}$ and $\{1, -2, +1\}$, ... ; iii) A third family of **ACs** around 8 T (related to states between 130 and 180 meV) involves only $n = 2$ levels. The first **ACs** being between $\{0, 2, -1\}$ and $\{1, 1, +1\}$, $\{1, 1, -1\}$ and $\{2, 0, +1\}$, $\{2, 0, -1\}$ and $\{2, -1, +1\}$, ... Although lowest **ACs** in the **QD** spectrum are caused by the selection rules of H_R, the presence of H_D^L and H_D^C, in the full H, displaces and regroup all states with same n value that contribute to the minigap near a fixed critical field value.

In Fig. 4 we simulate the cancellation of zero-field splittings even in the presence of all **SO** terms, what is reasonably obtained by taking an interfacial field dV/dz four times stronger than that one considered before (see Ref. [7], other parameters remained unchanged). This is equivalent to increase the influence of the Rashba term H_R. Notice, in Panels A and B, that not only the zero-field splittings are nearly vanished, but also the Zeeman splittings are

G4.6.4

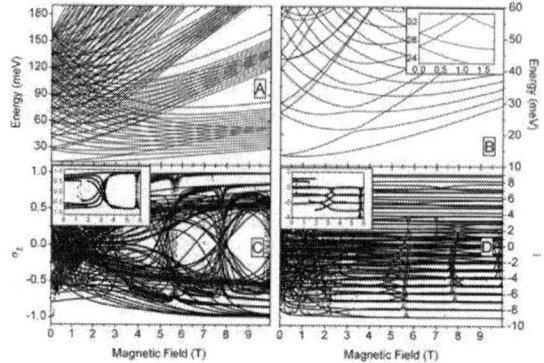

FIG. 3: Spectrum of full H, where H_R induces minigap regions that are shifted to higher fields by H_D^L. H_D^C has small influence but induces state mixtures. The zero-field splittings produced by H_{SIA}^D are dominated by those from H_D^L.

practically suppressed at low fields ($B_0 \lesssim 1.5$ T). At zero magnetic field, an energy shell structure identical to the pure H_0 and with the same level separation of 15 meV is formed at displaced energies. In the inset of Panel B one sees that the energy of $j = 3/2$ level is pushed near $j = 1/2$ level, going from 30 (in Fig.3 B) to 26.5 meV (in Fig.4 B). While the zero-field splittings nearly vanish the energy minigaps increases, as seen in Panel B. The rearrangement of electronic levels is so remarkable that **ACs** related to the cubic **BIA** term at 1.2 T become visible (Panel B and insets of Panels C and D). The minigaps at 33 (44) meV involves states $\{0, 1, -1\}$ and $\{0, -2, +1\}$ ($\{1, 0, -1\}$ and $\{0, -3, +1\}$). Even though the electronic levels are less disperse here than in Fig. 3, the SO-induced state mixture is much more intense, as can be seen in Panels C and D. Between 0 and 4 T, most of the **QD** levels have $|\sigma_Z| < 0.5$ and only the ground state has $\sigma_Z \simeq 0.7$. As mentioned before, the insets of Panels C and D show that a strong Rashba term enlarges the spin-flip region near B_C.

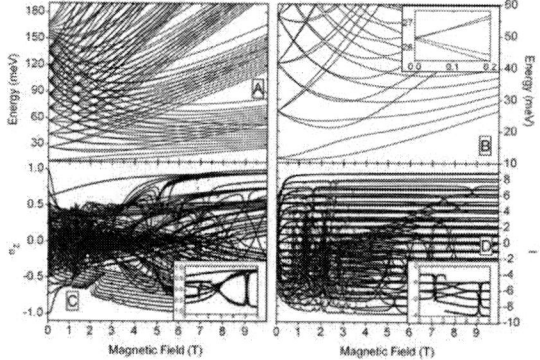

FIG. 4: Full H spectrum with four times stronger dV/dz. Notice the cancellation of zero-field and Zeeman splittings at low fields (A, B and inset). New **ACs** due to H_D^C selection rules occur near 1.2 T (B and insets in C and D). The lowest **AC** is shifted back to $B_C = 2.7$ T. Notice the enormous state mixture in C and, at zero field, most states are displaying $|\sigma_Z| < 0.5$.

One can further appreciate the intricate balance between **SO** terms by analyzing how some quantities are affected by changes on the lateral and vertical sizes, l_0 and z_0, or on Rashba field, dV/dz, as shown in Fig. 5. Curves with squares, circles and triangles refer to a **QD**, respectively, with parameters of Ref. [7], with z_0 doubled (smaller linear **BIA** contribution) and with four times stronger dV/dz, while the dotted curve, in middle Panel, indicates the B_C^0 field where the first **FD** level crossing occurs. The zero-field splitting (left Panel) for states $j = 3/2$ and $j = 1/2$ of the second shell is dominated by the linear **BIA** contribution for any value of l_0. An increase on z_0 strongly reduces

40

the splittings because the Dresselhaus contribution becomes weaker. The reduction is even more drastic by increasing dV/dz, which makes H_R larger and, thus, may cancel or suppress zero-field splittings produced by H_{SIA}^D.

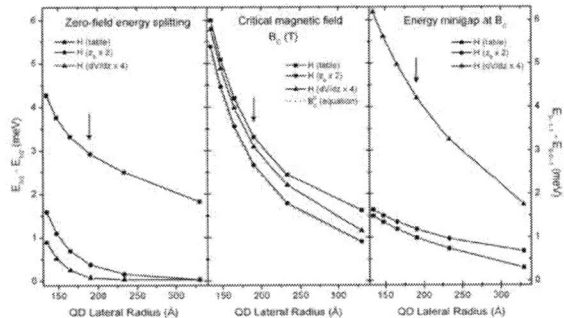

FIG. 5: Zero-field energy splittings for the states in the second energy shell (left Panel), critical magnetic fields where the first level **AC** (middle Panel) occurs, and energy minigaps opened at that AC (right Panel) as function of the **QD** lateral radius l_0. Meaning of square, circle and triangle curves are explained in text. Arrows at $l_0 = 190$ Å show the **QD** radius where the spectra from Figs. 1 to 4 were calculated.

The critical fields B_C, where **AC**s determined by H_R occur (middle Panel, for the lowest minigap between levels $\{0,0,-1\}$ and $\{0,-1,+1\}$), decrease with increasing **QD** size, once $B_C^0 \simeq \omega_0 \simeq 1/\sqrt{l_0}$. Its value is close to B_C^0 when **BIA** terms are not present and the inclusion of H_D^L shifts B_C to higher value. Increasing z_0 or dV/dz decreases B_C, once they will decrease the effects due to H_{BIA}. The value $B_C \simeq 2.1$ T, for $l_0 = 270$ Å ($\hbar\omega_0 = 7.5$ meV) is displaced to 1.8 (1.5 T) if dV/dz (z_0) is four times larger (doubled). This last value can be compared to reported $B_C \simeq 1.7$ T in Ref. [8], where **BIA** terms are absent. The small difference $\Delta B = 0.2$ T can be attributed to the inclusion of non-parabolicity effects. Anticrossings at such low fields may be interesting for applications due to easier access.

Finally, the minigap opened at B_C (right Panel) has their main origin in the H_R term, while the inclusion of H_{BIA} causes a substantial reduction. If the value of z_0 is doubled the splitting is enhanced slightly. A yet larger z_0 produces no significant changes. However, the splitting can be drastically enhanced by increasing the Rashba field as, for example, changing from 1 to 4.2 meV at $l_0 = 190$ Å, when interfacial field is increased four times. Measurement of those three quantities should yield important information on the relative strength of **SO** parameters α and γ.

After having studied the one-particle **QD** problem we show, in Fig. 6, the two-electron **QD** spectrum under magnetic field (parameters in Ref. [7]). On the construction of Slater determinant for two-particle states we use the 20 lowest one-particle orbitals ($|l| \le 3$ and $n \le 1$), which amounts to 190 possible two-particle states that can be labeled, in the absence of **SO** interactions, by the projections of orbital (M_L) and spin (M_S) total angular momenta. If no **SO** term is present in H (Panel A), we verified that the singlet ground state is located at 35 meV, while the first excited shell at zero-field is splitted by the exchange interaction, being composed by a triplet (at 47.5 meV) and a singlet (at 50 meV) states. At very small magnetic field ($\simeq 0.1$ T), the normal sequence of QD states is restored. For increasing energy and using the notation $\{M_L, M_S\}$, the ordering is: $\{0,0\}$ for the ground state, $\{-1,1\}$, $\{-1,0\}^T$, $\{-1,-1\}$, $\{1,1\}$, $\{1,0\}^T$, $\{1,-1\}$ for the first excited triplet (T), and $\{-1,0\}^S$, $\{1,0\}^S$ for the first excited singlet (S). The crossing between ground singlet and first excited triplet states occurs at $B_C^{0(2e)} = 2.1$ T.

Panel B shows the **QD** spectrum for full two-particle Hamiltonian, $H_{ee} + H$. We may identify some similar features to the single-particle case. For example, the linear Dresselhaus term almost destroys the energy shell structure at zero field by shifting level crossings and inducing new zero-field splittings, while the Rashba term introduces energy minigaps in the spectrum. Panel B shows details on the competition between Coulomb and spin-orbit interactions in narrow-gap cylindrical dots. Observe that the **SO** interaction, at zero field, acts against the direct Coulomb energy and, in a sense, favoring the exchange term. For example: i) The ground state is shifted back from 35 to 31 meV, which is close to the non-interacting value 30 meV; ii) The first excited shell states have energies from 43 to 47 meV, values even smaller than the non-interacting energy 45 meV. Other important feature in the first excited shell is the observation that the original triplet is broken on its three possible terms according to the projection of the total angular momentum, $M_J = M_L + M_S$. For increasing energy order, these terms are composed, at zero field, by the states $\{-1,1\}$ and $\{1,-1\}$ ($M_J = 0$), $\{-1,0\}^T$ and $\{1,0\}^T$ ($|M_J| = 1$), $\{-1,-1\}$ and $\{1,1\}$ ($|M_J| = 2$), in increasing

G4.6.6

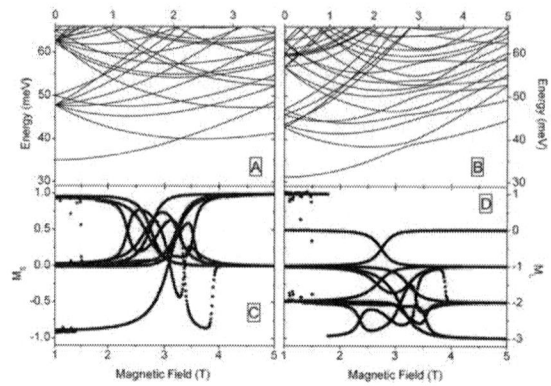

FIG. 6: Two-particle **QD** spectrum without (A) and with (B) all **SO** terms. It is seen that the **SO** energy acts against the direct and in favor of the exchange Coulomb energies. The first excited triplet, at zero field, is splitted according to the possible M_J values as explained in text. C (M_S) and D (M_L) show the lowest ACs as induced by H_R and shifted by H_D^L.

energy, while the ground ($\{0,0\}$, $M_J = 0$) and first excited ($\{-1,0\}^S$ and $\{1,0\}^S$, $|M_J| = 1$) singlets remain the same.

Panels C and D show the **SO**-induced mixing of those lowest states. The first **AC** at $B_C^{(2e)} = 2.7$ T involves states $\{0,0\}$ and $\{-1,1\}$ (H_R selection rule yields $\Delta M_L = \pm 1 = -\Delta M_S$), so that the difference between $B_C^{(2e)}$ and $B_C^{0(2e)}$ (~ 0.6 T) is basically the same between B_C and B_C^0 (~ 0.7 T) for the one-electron problem. This means that the shifting of B_C due to H_D^L is not altered by **QD** occupation, although B_C itself is decreased by increased occupation. Yet, the H_D^C selection rule becomes $\Delta M_L = \mp 3$ and $\Delta M_S = \pm 1$, and its first minigap between states $\{-2,1\}$ and $\{1,0\}^{S/T}$ would be visible if a higher dV/dz had been considered on the solution for the two-electron problem.

A very special difference between the one- and two-particle problems is that a strong intrinsic (no phonon-assisted) singlet-triplet transition (qubit) at low magnetic fields involving the ground state becomes possible in the two-electron case and, in principle, could be explored in implementations of quantum computing devices. As mentioned, the critical field is decreased by **QD** occupation (from $B_C = 3.3$ to $B_C^{(2e)} = 2.7$ T), and this reduction may be increased by decreasing the **QD** confinement energy. At these critical fields where the intrinsic state mixture is enhanced, the **SO**-induced spin relaxation rate (Γ) can be estimated from the minigap energy (Δ), as $\Gamma = \hbar/\Delta$. For the lowest **AC**, Δ values are taken from the right Panel of Fig. 5, from where one sees that Δ is completely changeable by the **QD** parameters and, consequently, the intrinsic rate Γ can be changed according to those parameters.

We showed that inclusion of all **SO** terms is essential in order to obtain a complete picture of the electronic structure of narrow-gap **QD**s, and discussed the role played by each **BIA** and **SIA** terms on **QD** spectra and on spin polarization of states. The combination of strong **SO** coupling in H_R and large g-factor introduces strong intrinsic mixtures and low excitations on the single-particle spectrum; the position of critical fields where minigaps occur is affected by H_D^L. We observed that the two-particle spectrum exhibits strong singlet-triplet coupling involving QD ground state at moderate fields, which may have significant consequences like possible use in qubits designs.

Work supported by FAPESP-Brazil, US DOE grant no. DE-FG02-91ER45334, and CMSS Program at OU.

---

[1] S. Datta, and B. Das, Appl. Phys. Lett. **56**, 665 (1990).

[2] Y.A. Bychkov, and E.I. Rashba, J. Phys. C **17**, 6039 (1984).

[3] D. Loss, and D.P. DiVincenzo, Phys. Rev. A **57**, 120 (1998); X. Hu, and S. Das Sarma, Phys. Rev. A **64**, 042312 (2001).

[4] G. Dresselhaus, Phys. Rev. **100**, 580 (1955).

[5] L. Jacak, A. Wojs, and P. Hawrylak, Quantum Dots (Springer, Berlin, 1998).

[6] C.F. Destefani, S.E. Ulloa, and G.E. Marques, Phys. Rev. B **69**, 125302 (2004); C.F. Destefani, Ph.D. Thesis at Universidade Federal de São Carlos, São Paulo, Brazil, unpublished.

[7] InSb parameters: $m = 0.014\ m_0$, $g = -51$, $\varepsilon = 16.5$, $a_B = 625$ Å, $\alpha = 500$ Å^2, $\gamma = 160$ eVÅ^3. Dot characteristics: $\hbar\omega_0 = 15$ meV ($l_0 = 190$ Å), $z_0 = 40$ Å, $dV/dz = -0.5$ meV/Å. Prefactors (meV) at zero B-field: $E_{SIA}^D = \alpha\hbar\omega_0/l_0^2 = 0.2$, $E_R = -(\alpha/\lambda)dV/dz = 1.3$, $E_D^C = \gamma/\lambda^3 = 0.02$, $E_D^L = \gamma\langle k_z^2\rangle/\lambda = 5.2$, and $E_{ee} = \hbar\Omega\lambda/a_B = 4.5$.

[8] T. Darnhofer, and U. Rössler, Phys. Rev. B **47**, 16020 (1993).

Mat. Res. Soc. Symp. Proc. Vol. 825E © 2004 Materials Research Society

Theory of Electrically Controlled Resonant Tunneling Spin Devices

David Z.-Y. Ting and Xavier Cartoixà [1,*]
Jet Propulsion Laboratory, California Institute of Technology,
Pasadena, CA 91109-8099, U.S.A.
[1]Department of Physics, University of Illinois at Urbana-Champaign,
Urbana, IL 61801, U.S.A.

ABSTRACT

We report device concepts that exploit spin-orbit coupling for creating spin polarized current sources using nonmagnetic semiconductor resonant tunneling heterostructures, without external magnetic fields. The resonant interband tunneling spin filter exploits large valence band spin-orbit interaction to provide strong spin selectivity. The bi-directional spin pump induces the simultaneous flow of oppositely spin-polarized current components in opposite directions through spin-dependent resonant tunneling. The efficiency of resonant tunneling spin devices can be improved when the effects of structural inversion asymmetry (SIA) and bulk inversion asymmetry (BIA) are combined properly, and incorporated into device design. The current spin polarizations of the proposed devices are electrically controllable, and potentially amenable to high-speed modulation. In principle, the electrically modulated spin-polarized current source could be integrated in optoelectronic devices for added functionality.

INTRODUCTION

An important component of semiconductor spintronics (spin-based electronics) research is the development of spin-polarized current sources [1]. One interesting approach uses non-magnetic semiconductor heterostructures, without external magnetic fields or optical excitation. The idea originated with the resonant tunneling spin filter proposed by Voskoboynikov et al. [2]. Subsequently, a number of new device concepts emerged, including the triple-barrier resonant tunneling diode (TB-RTD) [3], the asymmetric resonant interband tunneling diode (aRITD) [4,5], the bi-directional resonant tunneling spin pump [6], and the [110]-RITD [7]. In this paper we present an overview of some of these concepts, and discuss their prospects.

DEVICE CONCEPTS

We illustrate the concept of nonmagnetic heterostructure spin filters using the asymmetric resonant tunneling structure (aRTS) as an example. Quantized states in aRTS are spin-split by the Rashba effect [8]. Spin filtering is done by exploiting the phenomenon that the spin of a resonantly transmitted electron aligns with that of the quasibound state traversed [2, 9]. Figure 1(a) illustrates the properties of quasibound quantum well states in an aRTS. Rashba effect induced spin splitting in the lowest conduction band (cb1) states are shown in the left panel. The shaded disks in the right panel are $k_\parallel$-space representations of available quasibound states with energy below the Fermi level in the incident electron reservoir. These are the states that participate in resonant tunneling, and their spin directions determine the spin polarization of the transmitted electrons in the collector. When the spin directions of two spin-split subbands are

43

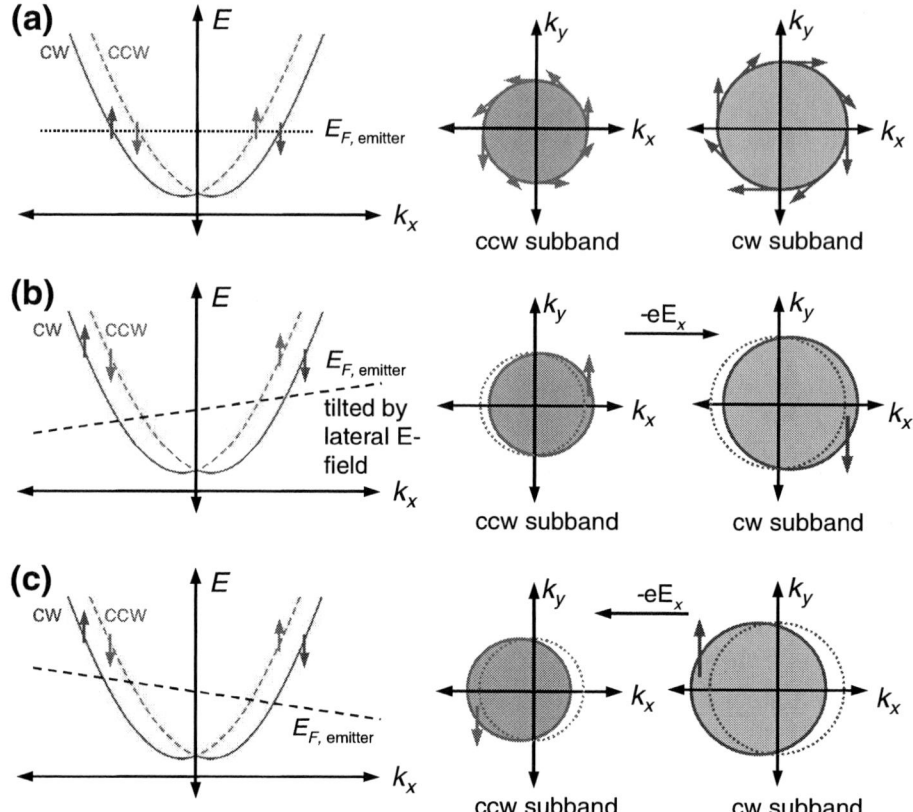

Figure 1. (a) Lowest conduction subband (cb1) in an asymmetric quantum well showing Rashba spin splitting. Available quasibound states in the cb1 with energy below the emitter Fermi level are shown as shaded disks in $\mathbf{k}_\parallel$-space. Arrows indicate spin directions. (b) Same as (a), but with the Fermi surface in the incidence electrode displaced by the application of a lateral E-field along the x direction. (c) Same as (b), but with the lateral E-field in the opposite direction, resulting in the opposite net spin polarization.

plotted along the disk perimeters, they appear in counter-clockwise (ccw) and clockwise (cw) pinwheel patterns. Thus we label the subbands "ccw" and "cw." Note that ccw and cw subband states at the same $\mathbf{k}_\parallel$ have opposite spins, and that states with opposite $\mathbf{k}_\parallel$ within a given subband have opposite spins. Thus, to achieve efficient spin filtering we must provide mechanisms for (1) preferentially selecting a particular spin-split subband, and (2) lateral momentum selection.

The subband filtering efficiency is defined as $\eta = (J_{ccw}-J_{cw})/(J_{ccw}+J_{cw})$, where J_{ccw} and J_{cw} are the resonant tunneling current density components associated with the ccw and cw subband tunneling, respectively. In the aRTS η is typically limited by cancellation between the ccw and cw subband contributions. The strategy for optimizing η is to increase spin splitting, thereby enlarging the difference between ccw and cw contributions. One way to strengthen the selectivity between the ccw and cw subbands is to use the spin-blockade mechanism proposed by Koga *et al.* [3]. Their proposed device consists of a triple-barrier structure containing back-to-back asymmetric wells coupled through a thin central tunneling barrier. The two quantum wells have opposite ordering of the ccw and cw subbands. Resonant tunneling is blocked unless the quasibound state spins in the two wells are aligned (spin blockade). Either ccw-ccw or cw-cw alignment can be selected by the biasing voltage. This technique has resulted in very high calculated subband filtering efficiency of $\eta > 99.9\%$ [3]. Alternativlely, we can improve spin filtering efficiency by exploiting the strong spin-dependent interband tunneling through hole states in asymmetric resonant interband tunneling diodes (aRITDs) [4]. The interband design uses large valence band spin-orbit interaction to provide strong spin selectivity, but does not leave the electrons in valence bands where spin relaxation is fast. Filtering efficiency is also enhanced by the reduction of tunneling through quasibound states near the zone center. Another interband device concept takes advantage of bulk inversion asymmetry (BIA) induced spin splitting in (110) devices to perform spin filtering [7].

Lateral momentum selection can be accomplished in several ways. Voskoboynikov *et al.* proposed using a small in-plane electric field in the source region of the aRTS to shift the incident electron distribution towards, say, the positive k_x side in k-space [2]. Figure 1(b) shows that a lateral E-field along the x direction would result in net $+y$ and $-y$ spin polarizations for resonant tunneling currents transmitted through the ccw and cw subbands, respectively. And since cw subband current contributions are larger, the total transmitted current yields a net $-y$ current spin polarization. Note that, as indicated in Fig. 1(c), reversing the direction of the lateral E-field causes the collector current to be spin polarized in the opposite direction. Thus, it is possible to modulate the spin polarization by changing the direction of the lateral E-field. A variant lateral E-field scheme, proposed by Chow and Moon [5], uses lateral side-gates fabricated on the mesa sidewalls of the resonant tunneling structure. The side gates are electrically isolated from the mesa itself, and do not induce a lateral current in the emitter region. An alternative for creating lateral momentum anisotropy was proposed by Datta and co-workers [3]. It makes use of a one-sided collector, which is placed on, for example, the positive x side to collect only electrons with positive k_x. It can be shown that for the one-sided collector geometry, the net current spin polarization, $P = (J_{+y}-J_{-y})/(J_{+y}+J_{-y})$, where J_{+y} and J_{-y} are resonant tunneling current density components spin polarized along the positive and negative y directions, respectively, is related to the subband filtering efficiency in a simple manner: $P = (2/\pi)\eta$ [10]. This limits P to a theoretical maximum of $(2/\pi) \approx 63.7\%$ for the one-sided collector geometry.

A bi-directional spin-pump [6] is similar in structure to the resonant tunneling spin filter. In the spin pump, we do not intentionally bias the device along the growth (z) direction, but apply a small lateral E-field in the emitter region only. Consider a spin pump structure where the emitter and the collector are made from the same material and doped to equal carrier concentrations. The application of an in-plane E-field along the x direction displaces the emitter Fermi surface. It creates an excess of carriers on the $+k_x$ side, which can tunnel to the collector, and a deficit of carriers on the $-k_x$ side, which becomes available to receive electrons tunneling back from the collector. Assuming the spin filter structure is designed such that resonant tunneling through the cw states dominates over the ccw states at zero bias, then resonantly transmitted electrons on the $+k_x$ and $-k_x$ sides will be spin polarized along the $-y$ and $+y$ directions, respectively. This results in a forward (emitter to collector) electron (particle) current with $-y$ spin polarization, and a backward current with $+y$ spin polarization. The bi-directional spin pump induces the simultaneous flow of oppositely spin-polarized current components in opposite directions, and can thus generate significant levels of spin current with very little net electrical current across the tunnel structure, a condition characterized by a greater-than-unity current spin polarization. Incidentally, the accumulation of spins in the bi-directional spin pump is reminiscent of the spin Hall effect [11], and may thus be analyzed in that framework.

The resonant tunneling spin filter and spin pump concepts were developed to exploit the Rashba effect [8], which is a consequence of spin–orbit interaction and the presence of structural inversion asymmetry (SIA). The effect on spins due to the presence of bulk inversion asymmetry (BIA) in zincblende semiconductors can also be exploited for spin device applications. It can be shown that the efficiency of nonmagnetic resonant tunneling spin devices can be improved significantly when SIA and BIA effects are combined properly [10]. The design changes required to take advantage of this improvement are minimal: we only need to be more specific in selecting the direction of one-sided collectors or the lateral E-field [10].

MODELING RESULTS

In this section we present spin filter modeling results, calculated using the multiband quantum transmitting boundary method [12], on an InAs-AlSb-InAs-GaSb-AlSb-InAs asymmetric resonant interband tunneling diode (aRITD). Band structure is described by the effective bond orbital model [13], using typical 8-band $k \cdot p$ material parameters taken from the literature. The left panel of Fig. 2 shows the results for the case with applied lateral E-field in the emitter. Three different E-field strengths (E_x=0, 33, and 67 V/cm) are used in this calculation. Spin polarized current density components J_{+y} and J_{-y} are shown in the top panel, and the current spin polarization in the bottom. As expected, no spin polarization is found for E_x=0. But significant spin polarization is found for E_x=33 and 67 V/cm for applied biases under 0.05 V. The reason that there is no current spin polarization at the peak of the J-V curves near 0.11 V is because the peak is dominated by light-hole 1 (lh1) resonant tunneling. The lh1 tunneling process favors contributions from the zone-center, where Rashba spin splitting

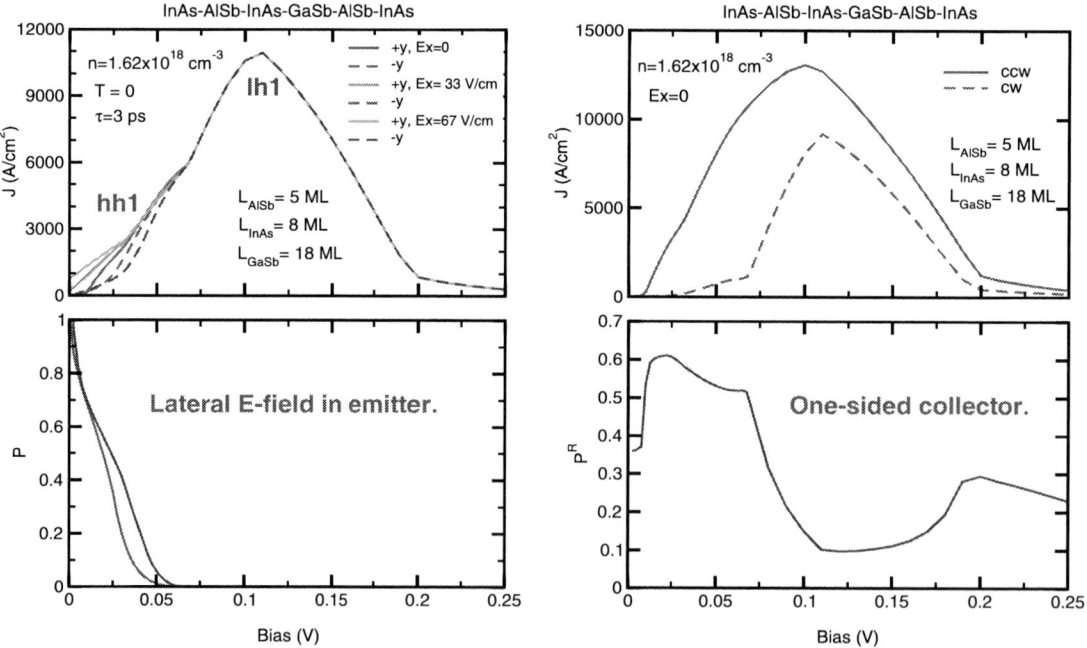

Figure 2. Calculated current density components (top) and current polarization efficiencies (bottom) in an aRITD, as functions of applied bias. The left and right panels show the results for the lateral E-field and the one-sided collector geometries, respectively.

vanishes. At below 0.05V, heavy-hole 1 (hh1) tunneling dominates. Here the away-from-zone-center resonant tunneling processes, which show strong spin dependence, are favored. Note that as the applied bias approaches 0, the $E_x=33$ and 67 V/cm cases show collector current spin polarization greater than 1. This is a manifestation of the bi-directional spin pump effect [6]. The right panel of Fig. 2 shows the computed subband current densities J_{ccw} and J_{cw}, and one-sided collector current spin polarization P as functions of applied bias for the same aRITD. It shows that the aRITD can be a highly efficient spin filter, with P approaching the theoretical maximum of 63.7% at relatively large current density levels.

DISCUSSIONS

The concept of non-magnetic heterostructure spin filters is still being refined. Here we point out a few areas that are under development. A common assumption used in the resonant tunneling spin filter models thus far is coherent tunneling. This preserves the lateral momentum anisotropy during the resonant tunneling process, so that net spin polarization can be obtained in the collector. Coherent tunneling is more likely to occur when resonant tunneling lifetimes are shorter than typical scattering times in the quantum well. For the aRITD structure studied in Fig. 2, the typical tunneling lifetime is on the order of 10 ps. Studies in designing spin-dependent tunneling devices with shorter tunneling lifetimes are ongoing. There has been indirect evidence for the demonstration of coherent tunneling in large-current density resonant tunneling devices [14]. It should be noted that coherent tunneling is a sufficient, but not a necessary requirement for the resonant tunneling spin filter. More detailed studies of the dynamics of spin transport in resonant tunneling spin filters are needed for better understanding.

Another issue being study currently is the effect of higher order k terms. We have shown that BIA effects can be used to enhance spin filtering efficiencies using a model Hamiltonian that describes linear-in-k spin splitting for states near the zone center [10]. The recent work of Winkler points out that the quantum well state spin directions can be more complex when higher-order k terms are included [15], and we expect the higher-order k terms to have important effects for transport as well. These effects are being studied using a band structure method that incorporates BIA effects in the eight-band effective bond orbital model (BIA-EBOM) [16].

A salient feature of the non-magnetic semiconductor heterostructure spin devices is that they do not need magnetic elements (e.g. Mn), and are compatible with conventional semiconductor growth technology. This means that they can be grown, for instance, in the same molecular beam epitaxy machine that is used to grow lasers, detectors, or transistors. However, the need for lateral momentum selection means that they require sophisticated device design and processing technology, such as those used in the fabrication of the split side-gated resonant interband tunneling devices [17].

A resonant tunneling spin filter implemented in the lateral E-field geometry (including split side gate) offers us the ability to control the current spin polarizations electrically, as illustrated in Fig. 1(b) and 1(c). Therefore it is potentially amenable to high-speed spin modulation through electrical means, such as side-gate voltage modulation. A possible application for a spin polarized current source, capable of being modulated electrically at high speeds, is the electrically pumped spin VECSL [18]. In the "spin laser" light polarization follows modulation in injected carrier spin, and can carry information in both light polarization and intensity.

SUMMARY

We reported progress on the development of device concepts that exploit spin-orbit coupling for creating spin polarized current sources using nonmagnetic semiconductor resonant tunneling heterostructures, without external magnetic fields. These spin polarized current source are potentially capable of being modulated electrically at high speeds, and could be useful for spintronics and spin-optoelectronics applications.

ACKNOWLEDGMENTS

The authors thank O. Voskoboynikov, A. T. Hunter, D. L. Smith, D. H. Chow, J. S. Moon, T. C. McGill, T. F. Boggess, J. N. Schulman, P. Vogl, T. Koga and Y.-C. Chang for discussions. This work was sponsored by the DARPA SpinS Program through HRL Laboratories. A part of this work was carried out at the Jet Propulsion Laboratory, California Institute of Technology, through an agreement with the National Aeronautics and Space Administration.

* Present address: Computational Research Division, Lawrence Berkeley National Laboratory, Berkeley, CA 94720, USA

REFERENCES

1. S. A. Wolf, D. D. Awschalom, R. A. Buhrman, J. M. Daughton, S. von Molnar, M. L. Roukes, A. Y. Chtchelkanova, and D. M. Treger, *Science* **294** 1488 (2001).
2. A. Voskoboynikov, S. S. Lin, C. P. Lee, and O. Tretyak, *J. Appl. Phys.* **87**, 387 (2000).
3. T. Koga, J. Nitta J, H. Takayanagi, and S. Datta, *Phys. Rev. Lett.* **88** 12661 (2002).
4. D. Z-Y. Ting and X. Cartoixà, *Appl. Phys. Lett.* **81**, 4198 (2002).
5. D. Z-Y. Ting, X. Cartoixà, D. H. Chow, J. S. Moon, D. L. Smith, T. C. McGill and J. N. Schulman, *IEEE Proceedings* **91** 741 (2003).
6. D. Z.-Y. Ting and X. Cartoixà, *Appl. Phys. Lett.* **83**, 1391 (2003).
7. K. C. Hall, W. H. Lau, K. Gundogdu, M. E. Flatte and T. F. Boggess, *Appl. Phys. Lett.* **83**, 2937 (2003).
8. Y. A. Bychkov and E. I. Rashba, *J. Phys. C - Solid State Phys.* **17**, 6039 (1984).
9. E. A. de Andrada e Silva and G. C. La Rocca, *Phys. Rev. B* **59**, 15583 (1999).
10. D. Z.-Y. Ting and X. Cartoixà, *Phys. Rev. B* **68**, 235320 (2003).
11. J. E. Hirsch, *Phys. Rev. Lett.* **83**, 1834 (1999).
12. D. Z.-Y. Ting, E. T. Yu, and T. C. McGill, *Phys. Rev. B* **45**, 3583 (1992).
13. Y.-C. Chang, *Phys. Rev. B* **37**, 8215 (1988).
14. R. C. Bowen, G. Klimeck, R. K. Lake, W. R. Frensley, and T. Moise, *J. Appl. Phys.* **81**, 3207 (1997); W. R. Frensley, private communications.
15. R. Winkler, *Phys. Rev. B* **69**, 045317 (2004).
16. X. Cartoixà, D. Z.-Y. Ting, and T. C. McGill, *Phys. Rev. B* **68**, 235319 (2003).
17. J. S. Moon et al. (unpublished).
18. M. Oestreich, M. Bender, J. Hubner, D. Hagele, W. W. Ruhle, T. Hartmann, P. J. Klar, W. Heimbrodt, M. Lampalzer, K. Volz, W. Stolz, *Semicond. Sci. Technol.* **17**, 285 (2002).

Growth of mirror-like $Zn_{1-x}Mn_xO$ diluted magnetic semiconductor thin films by r.f. magnetron sputtering method

Sejoon Lee[1], Hye Sung Lee[1], and Deuk Young Kim[1*]
[1] Quantum-functional Semiconductor Research Center, Dongguk University
3-26 Phil-dong, Chung-gu, Seoul 100-715, KOREA

ABSTRACT

The $Zn_{1-x}Mn_xO$ thin films were grown on Al_2O_3 (0001) substrates by an r.f. magnetron sputtering method. The film grown with employing buffer layer shows mirror-like surface, while the film grown without buffer layer shows the columnar-structured configuration. The mirror-like $Zn_{0.93}Mn_{0.07}O$ thin films have the single crystalline phase with (000ℓ) orientation normal to the substrate surface and show the UV emission originated from the near band-edge-emission for the measurements of x-ray diffraction and photoluminescence, respectively. The mirror-like $Zn_{0.93}Mn_{0.07}O$ film clearly showed a hysteresis loop, which is obvious evidence of ferromagnetism, and the Curie temperature was determined to be 68 K for the characterization of the temperature-dependent magnetization.

INTRODUCTION

Diluted magnetic semiconductors (DMSs) are most favorable materials for spintronic device applications, on account of providing new functionality for the semiconductor devices using both charge and spin degree of the freedom [1-2]. Possible spintronic devices are spin-valve transistors, spin light-emitting diodes, non-volatile storage and logic devices, and solar sells [3-5]. For the realization of those applications, the DMSs should show characteristics of ferromagnetism with the high Curie temperature (T_C) as well as the single crystalline phase. In addition, the formation of high quality diluted magnetic semiconductor (DMS) thin films is an important issue because heteroepitaxial films are required for applications of sophisticated spintronic devices. Since Dietl predicted that the Mn-doped ZnO would show the ferromagnetic behavior with the T_C of 300 K [6], great deal of studies on the $Zn_{1-x}Mn_xO$ DMSs have been progressed.

Recently, Norton [7] produced the Mn-implanted ZnO:Sn single crystal with the T_C of 250 K. However, Mn-implanted ZnO:Sn DMSs are not the thin film but just bulk single crystal even though they have fairly high T_C. In association with a study on ferromagnetic $Zn_{1-x}Mn_xO$ thin films, the report on high quality $Zn_{1-x}Mn_xO$ thin films with high T_C ferromagnetism is still poor,

---

[*] Corresponding Author. E-mail Address: dykim@dongguk.edu

although a few studies [8-9] have been reported. Thus the study on the growth and the characterization of $Zn_{1-x}Mn_xO$ thin film is keenly required to realize the spintronic device applications. In this paper, we report on material properties of mirror-like $Zn_{1-x}Mn_xO$ thin films grown by the r.f. magnetron sputtering deposition.

EXPERIMETAL DETAILS

$Zn_{1-x}Mn_xO$ thin films were grown on Al_2O_3 (0001) substrates using the r.f. magnetron sputtering deposition method. The sputtering target of $Zn_{1-x}Mn_xO$ (Mn: 5 wt%) used in this work was made using a standard ceramic synthesis method. In our growth process, the mixture of gases (Ar 12 sccm : O_2 12 sccm) was flowed, the r.f. power of 120W was applied, and the working pressure was kept to be 10 mTorr. The substrate temperatures are maintained at 200°C and 450°C during buffer and main film growth, respectively. After the growth of 40 nm-thick buffer layer, the 500 nm-thick main film was subsequently grown on buffer layer without break. The measurement of scanning electron microscopy (SEM) using a FE SEM XL-30 system was performed to monitor the cross-sectional and surface images of films, and the Mn content was measured by *in-situ* energy dispersive x-ray (EDX) spectroscopy using an EDAX NEW XL-30 system. The surface topology of buffer layer was measured by atomic force microscopy (AFM) utilizing a Digital Instruments Nanoscope IIIa system. The measurement of x-ray diffraction (XRD) using a Cu Kα source from Bede D3 system was carried out to evaluate the crystal properties of $Zn_{1-x}Mn_xO$ thin film. Temperature-dependent photoluminescence (PL) measurements were employed to characterize the optical properties of the mirror-like $Zn_{1-x}Mn_xO$ thin films. The magnetic properties of $Zn_{1-x}Mn_xO$ thin film were measured by a superconducting quantum interference device (SQUID) magnetometer using a Quantum Design MPMSXL system.

RESULTS AND DISCUSSION

The surface morphology and the cross-sectional aspect of $Zn_{1-x}Mn_xO$ films were monitored by SEM measurements. Figure 1 shows surface, cross-sectional SEM images, and schematic views of $Zn_{1-x}Mn_xO$ films grown at 450°C without and with employing the buffer layer, and the AFM image of buffer layer grown at 200°C. The film grown with 40 nm-thick buffer layer shows the mirror-like configuration as shown in Fig. 1(d) and 1(e), while the film grown without buffer layer shows the columnar-structured configuration as shown in Fig. 1(a) and 1(b). It is considered that the amorphous buffer layer densely-packed with small columns may act as the strain-relaxation layer as shown in figures of 1(c) and 1(f) of schematic views of columnar and mirror-like $Zn_{1-x}Mn_xO$ cross-sections. The root mean square value of lateral grain size of the buffer layer shown in Fig. 1(g) is 37 nm approximately, and the formation mechanism of mirror-like $Zn_{1-x}Mn_xO$ thin films might be considered to be a result from the reduction of nucleation

G5.1.3

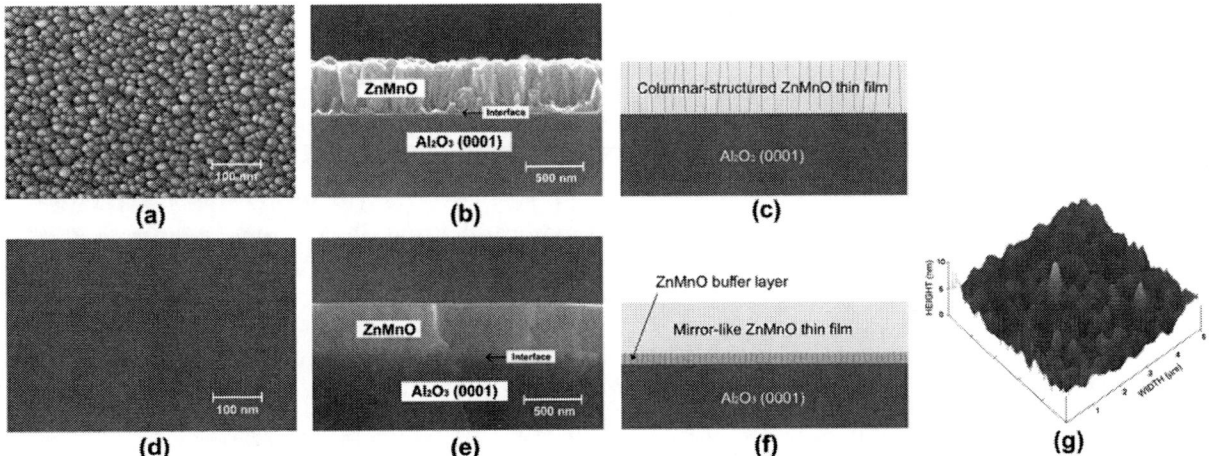

Figure 1. The SEM images of $Zn_{1-x}Mn_xO$ thin films and the AFM image of buffer layer grown by r.f. magnetron sputtering method. The (a) surface and (b) cross-sectional views are SEM images of the films grown at 450 ℃ without buffer layer, and the (d) surface and (e) cross-sectional views are SEM images of the films grown at 450 ℃ with employing a 40 nm-thick buffer layer. Figures of (c) and (f) show the cross-sectional schematics of (b) and (e), respectively. The (g) AFM image shows the surface topology of buffer layer grown on Al_2O_3 (0001) at 200 ℃.

sites for column growth precursors, and it is to be published elsewhere in more detail. The Mn contents were measured to be x=0.07 by the *in-situ* EDX measurement.

Figure 2 shows XRD spectrum of mirror-like $Zn_{0.93}Mn_{0.07}O$ films. The result from θ-2θ scan shows the peaks of $Zn_{0.93}Mn_{0.07}O$ (0002) and $Zn_{0.93}Mn_{0.07}O$ (0004), and the observation of only (000ℓ) without any secondary phase indicates that the $Zn_{0.93}Mn_{0.07}O$ film with employing buffer layer is single crystal grown with c-axis preference based on wurtzite structure. An inset in Fig. 2

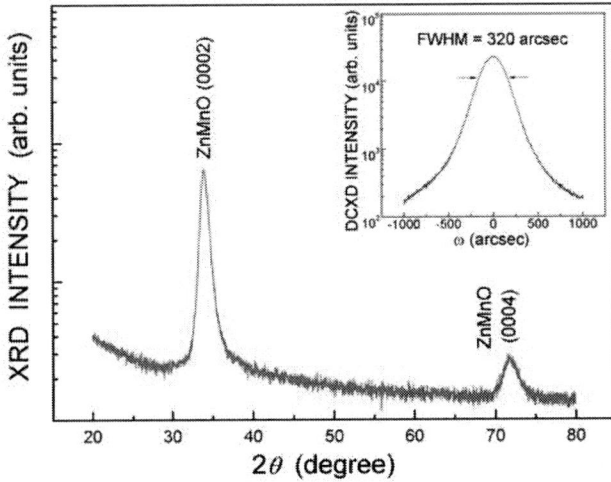

Figure 2. The XRD pattern of the mirror-like $Zn_{0.93}Mn_{0.07}O$ thin film. The inset shows double crystal rocking curve of $Zn_{0.93}Mn_{0.07}O$ (0002) peak.

shows double crystal rocking curve of $Zn_{0.93}Mn_{0.07}O$ (0002) peak. The full width at half maximum (FWHM) of diffraction angle is 320 arcsec (0.08°), it is comparable to the value of 0.04° for ZnO host material reported by An [10], therefore our mirror-like $Zn_{0.93}Mn_{0.07}O$ thin film is considered as the fairly good crystalline semiconductor without any precipitate and/or segregation. The lattice constants of c ~ 5.24 Å and a ~ 3.25Å for mirror-like $Zn_{0.93}Mn_{0.07}O$ thin films were determined from XRD measurements. It may reflect that the Mn ions were incorporated as much as 7%, because those values are similar to the result of Fukumura's work [11].

The temperature-dependent PL measurements were carried out in order to evaluate the optical properties related to the crystal quality of the mirror-like $Zn_{0.93}Mn_{0.07}O$ thin film. As shown in Fig. 3, the near band-edge-emission peak at 391 nm and deep level traps-related peak around 500 nm ~ 550 nm [12] were observed at 300 K. As the temperature decreases, the UV emission peak is split by two peaks at 386 nm and 404 nm those are expected to be the band-edge emission and the zinc vacancy-related emission, respectively. Since it is well know that when the Mn^{2+} ion substitute into the Zn^{2+} site Zn-related vacancy occurs, thus the peak at 404 nm might be related to the incorporated Mn components. In addition, there is no considerable change in peak intensity with decreasing temperatures, and it might be attributed to high exciton binding energy (60 meV). The results of PL measurements reflect that the mirror-like $Zn_{0.93}Mn_{0.07}O$ film is high quality crystal with semiconducting property.

The magnetic properties of the mirror-like $Zn_{0.93}Mn_{0.07}O$ thin film were confirmed using SQUID measurements. As shown in Fig. 4, the magnetization curve as a function of applied magnetic field shows a sharp hysteresis loop indicative of ferromagnetism. The remanent magnetization is 4.45 emu/cm^3 and the coercive magnetization is 240 Oe. An inset in Fig. 4 shows temperature-dependent magnetization, the ferromagnetic behavior of mirror-like $Zn_{0.93}Mn_{0.07}O$ thin film is kept to 68 K. The easy axis of our samples was observed at the in-plane direction, and it is due to the compressive strain-related anisotropy. Because the

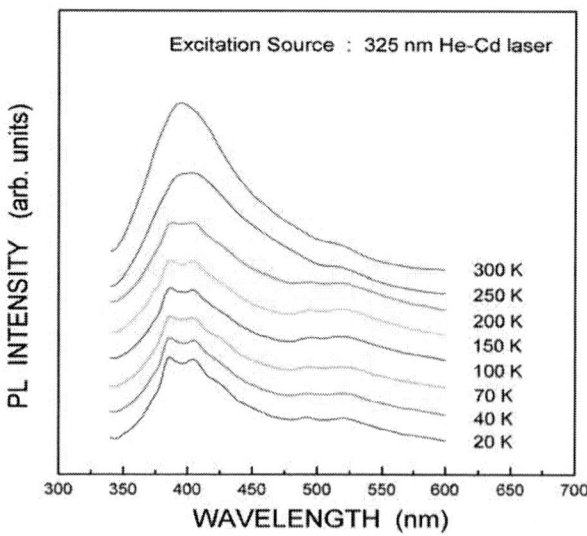

Figure 3. The temperature-dependent PL spectra of the mirror-like $Zn_{0.93}Mn_{0.07}O$ thin film.

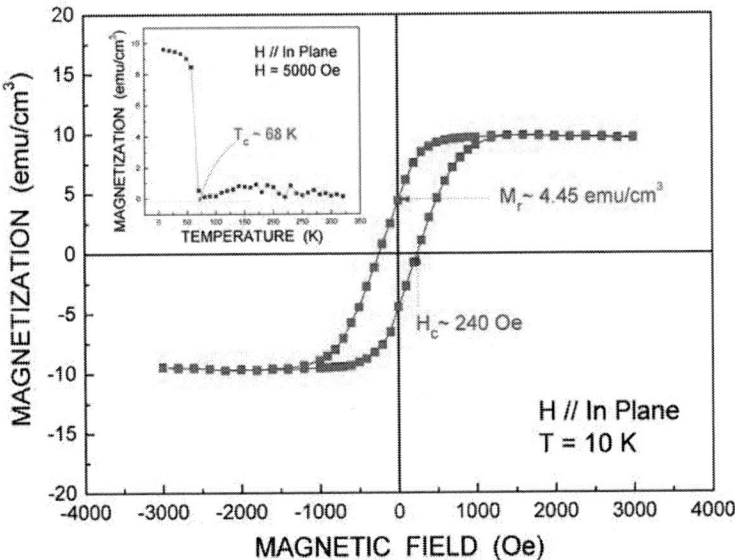

Figure 4. The curve of magnetization for the mirror-like $Zn_{0.93}Mn_{0.07}O$ thin film as a function of applied magnetic field. The inset shows the curve of magnetization for the mirror-like $Zn_{0.93}Mn_{0.07}O$ thin film as a function of temperature.

compressive strain at the film having hexagonal lattice structure is caused by preference of c-axis orientation, the observed easy axis reflect the mirror-like $Zn_{0.93}Mn_{0.07}O$ thin film is epitaxially grown with high crystal quality.

CONCLUSIONS

The mirror-like $Zn_{0.93}Mn_{0.07}O$ diluted magnetic semiconductor thin films on Al_2O_3 (0001) substrate were grown by the r.f. magnetron sputtering deposition method. The film grown with employing 40 nm-thick buffer layer has mirror-like surface and show single crystalline phase oriented with (000ℓ). The UV emission originated form the near band-edge-emission for the mirror-like $Zn_{0.93}Mn_{0.07}O$ thin film was observed for the PL measurements. From the results of magnetization measurements, the mirror-like $Zn_{0.93}Mn_{0.07}O$ thin film is obvious ferromagnetic semiconductor with the T_C of 68 K. These results experimentally suggest that high quality $Zn_{1-x}Mn_xO$ thin films can be effectively formed by the r.f. magnetron sputtering deposition method with employing strain-relaxation layer.

ACKNOWLEDGEMENT

This work was supported by the Korean Science and Engineering Foundation through the

Quantum-functional Semiconductor Research Center at Dongguk University.

REFERENCES

1. H. Ohno, Science **281**, 951 (1998).
2. S. A. Wolf, D. D. Awschalom, R. A. Buhrman, J. M. Daughton, S. von Molnár, M. L. Roukes, A. Y. Chtchelkanova, and D. M. Treger, Science **294**, 1488 (2001).
3. D. M. Bagnall, Y.F. Chen, Z. Zhu, T. Yao, S. Koyama, M.Y. Shen, T. Goto, Appl. Phys. Lett. **70,** 230 (1997).
4. M. H. Huang, S. Mao, H. Feick, H. Yan, Y. Wu, H. Kind, E. Weber, R. Russo, P. Yang, Science **292,** 897 (2001).
5. M. T. Bj.ork, B.J. Ohlsson, T. Sass, A.I. Persson, C. Thelander, M.H. Magnusson, K. Deppert, L.R. Wallenberg, L. Samuelson, Appl. Phys. Lett. **80,** 1058 (2002).
6. T. Dietl, H. Ohno, F. Matsukura, J. Cibert, and D. Ferrand, Science **287**, 1019 (2000).
7. D. P. Norton, S. J. Pearton, A. F. Hebard, N. Theodoropoulou, L. A. Boatner, and R. G. Wilson, Appl. Phys. Lett. **82**, 239 (2003).
8. S. W. Jung, S-J. An, G. C. Yi, C. U. Jung, S-I. Lee, and S. Cho, Appl. Phys. Lett. **80**, 4561 (2002).
9. D. S. Kim, S. Lee, C. Min, H. –M. Kim, S. U. Yuldashev, T. W. Kang, D. Y. Kim, and T. W. Kim, Jpn. J. Appl. Phys. **42**, 7217 (2003).
10. S.-J. An, W.I. Park, G.-C.Yi, and S. Cho, Appl. Phys. A **74**, 509 (2002).
11. T. Fukumura, Z. Jin, A. Ohtomo, H. Koinuma, and Kawasaki, Appl. Phys. Lett. **75**, 3366 (1999).
12. J. Wang, G. Du, Y. Zhang, B. Zhao, X. Yang, and D. Liu, J. Crystal Growth **263**, 269 (2004).

Mat. Res. Soc. Symp. Proc. Vol. 825E © 2004 Materials Research Society

Electronic structure of the diluted magnetic semiconductors Pb$_{1-x}$Sn$_x$Te:Yb

Evgenii Skipetrov, Elena Zvereva, Olga Volkova[1], Alexander Golubev[1] and Vasiliy Slyn'ko[2]
Low Temperature Physics Department, Faculty of Physics, Moscow State University,
Leninskie Gory, 119992 Moscow, RUSSIA.
[1]Faculty of Materials Science, Moscow State University,
Leninskie Gory, 119992 Moscow, RUSSIA.
[2]Institute of Material Science Problems,
Chernovtsy, 274001, UKRAINE.

ABSTRACT

The galvanomagnetic effects in the new diluted magnetic semiconductors Pb$_{1-x}$Sn$_x$Te:Yb were studied to determine the parameters of the electronic structure and to elucidate its influence on the magnetic properties. It was found that the temperature dependencies of the resistivity ρ and the Hall coefficient R_H have a "metallic" character, however the R_H changes in anomalous manner: its value increases more than by order of magnitude and then passes through maximum with increasing the temperature. Upon an increase of the ytterbium concentration the hole concentration decreases by more than order of magnitude. The results were explained assuming a formation of deep ytterbium-induced defect level in the valence band of the alloys, which moves up to its top with increasing the ytterbium concentration and pins the Fermi level within the valence band. The energy position of the Fermi level was calculated in the frame of two-band dispersion law and used to determine the position of Yb level in the alloys. The diagram of the charge carrier energy spectrum under varying the alloy composition was built.

INTRODUCTION

It has been known that doping of narrow-gap A^{4}B^6 semiconductors with mixed valence impurities results in a radical modification of the energy spectrum of charge carriers due to a formation of impurity induced deep defect levels [1,2]. It has been also found that some of such impurities transform the original materials into the diluted magnetic semiconductors, whose specific feature is that the magnetic activity of impurity ions is directly connected with their charge activity [3,4].

Recent experiments on ytterbium doped Pb$_{1-x}$Sn$_x$Te alloys revealed a paramagnetic behavior due to the presence of magnetically active Yb^{3+} ions lacking an electron on inside $4f$ electronic shell [5]. The amplitude of paramagnetic response increased with increasing the concentration of ytterbium introduced. However the dependence of magnetic Yb^{3+} ions concentration on the total amount of ytterbium was unusual: up to a certain critical value of impurity content the concentration of magnetic centers was approximately zero and then it rapidly grew.

At the same time the investigations on ytterbium doped Pb$_{1-x}$Ge$_x$Te alloys have shown that the magnetic properties of these alloys are strongly correlated with peculiarities of the electronic structure [6]. Located for these alloys slightly above the valence band top the Yb level moved up to the gap upon an increase of Ge or Yb content in the alloys and redistribution of the charge carriers between localized level and allowed bands occurred [7]. Under these conditions the magnetic properties of the alloys were determined not only by the Yb concentration but also by the occupancy of Yb-induced level and hence by the mutual arrangement of valence and Yb bands. By analogy we can assume that similarly to the case of the $Pb_{1-x}Ge_xTe$:Yb alloys the dependence of magnetic Yb^{3+} ions concentration on the total amount of ytterbium in the $Pb_{1-x}Sn_xTe$:Yb alloys will be governed by features of the electronic structure of these alloys.

Thus the main aims of this work was to establish the connection between the parameters of electronic structure and magnetic properties of the $Pb_{1-x}Sn_xTe$:Yb alloys and to build the energy level diagram under varying the matrix composition and Yb content.

EXPERIMENTAL DETAILS

Single crystals of p-type $Pb_{1-x}Sn_xTe$:Yb ($0.14 \leq x \leq 0.20$, C_{Yb}=0.1-1.6 mol%) were cut from the original boule grown by a modified Bridgman method. Ytterbium and tin were distributed along the boule in opposite to each other manner. Their concentrations were determined from the initial amount of substances in the furnace charge, taking into account the distribution of tin and ytterbium during the growth process according to the exponential law [8]. Table I summarizes the main parameters of the samples at T=4.2 K. For each sample the temperature dependencies of resistivity and Hall coefficient were measured by four probes technique ($B \leq 0.1$ T, $4.2 \leq T \leq 300$ K).

Table I. Parameters of the $Pb_{1-x}Sn_xTe$:Yb samples at T=4.2 K.

Sample	Tin content	Ytterbium content (mol%)	Resistivity (Ω cm)	Hall coefficient (cm^3/C)	Hole concentration (cm^{-3})
01	0.20	0.11	$1.2 \cdot 10^{-4}$	$9.8 \cdot 10^{-2}$	$6.4 \cdot 10^{19}$
03	0.18	0.12	$1.2 \cdot 10^{-4}$	$1.2 \cdot 10^{-1}$	$5.1 \cdot 10^{19}$
07	0.16	0.16	$1.5 \cdot 10^{-4}$	$1.8 \cdot 10^{-1}$	$3.4 \cdot 10^{19}$
11	0.16	0.21	$1.3 \cdot 10^{-4}$	$2.0 \cdot 10^{-1}$	$3.2 \cdot 10^{19}$
13	0.15	0.25	$1.4 \cdot 10^{-4}$	$2.3 \cdot 10^{-1}$	$2.8 \cdot 10^{19}$
15	0.15	0.31	$1.8 \cdot 10^{-4}$	$2.4 \cdot 10^{-1}$	$2.6 \cdot 10^{19}$
17	0.15	0.41	$1.9 \cdot 10^{-4}$	$2.1 \cdot 10^{-1}$	$3.0 \cdot 10^{19}$
21	0.14	0.90	$2.9 \cdot 10^{-4}$	$2.6 \cdot 10^{-1}$	$2.4 \cdot 10^{19}$
23	0.14	1.60	$1.5 \cdot 10^{-3}$	$6.1 \cdot 10^{-1}$	$1.0 \cdot 10^{19}$

RESULTS AND DISCUSSION

It was found that the temperature dependencies of resistivity ρ and Hall coefficient R_H have a metallic character over the whole temperature range studied (Fig. 1). However the Hall coefficient changes in anomalous manner (Fig. 1b): the value of R_H increases more than by order of magnitude with increasing the temperature. This is obviously indicative of the Fermi level (FL) pinning by the impurity ytterbium level, located within the valence band.

With increasing the ytterbium content C_{Yb} in the samples the resistivity and the Hall coefficient noticeably increase but have non-monotonous character: at certain impurity concentration C_{Yb} it undergoes a sudden leap and increases faster afterwards (Fig. 2). So we can

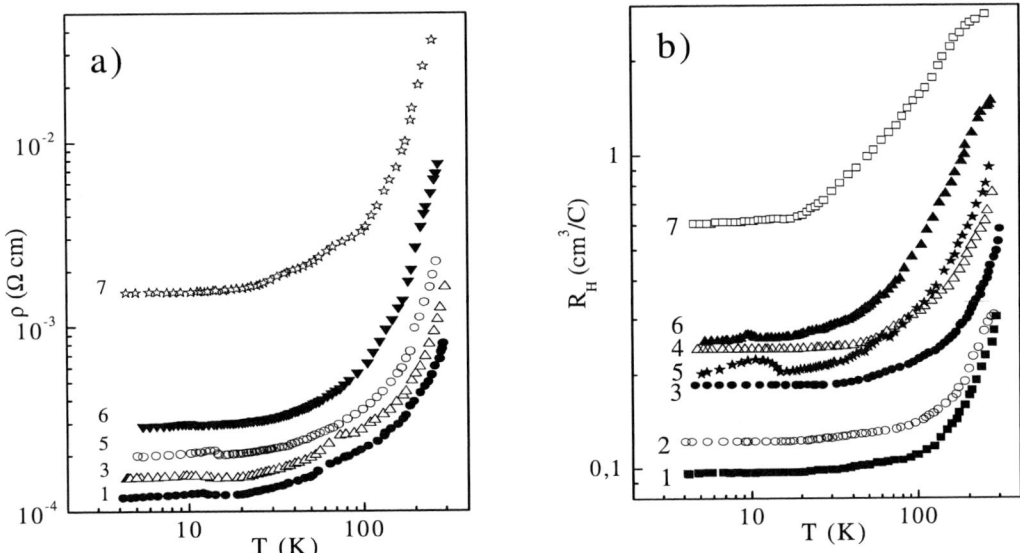

Figure 1. Temperature dependencies of the resistivity (a) and the Hall coefficient (b) in $Pb_{1-x}Sn_xTe:Yb$ alloys. Samples: 1 – 01, 2 – 03, 3 – 07, 4 – 15, 5 – 17, 6 – 21, 7 – 23.

conditionally divide the samples studied into two groups: with low Yb content (marked by black circles before leap in Fig. 2) and with high Yb content (open circles in Fig. 2). It is important to note that the magnetic measurements performed earlier for these samples [5] have shown that the samples of first group are nonmagnetic, while the samples of second group are paramagnetic ones and the concentration of magnetic centers increases rapidly with increasing C_{Yb}.

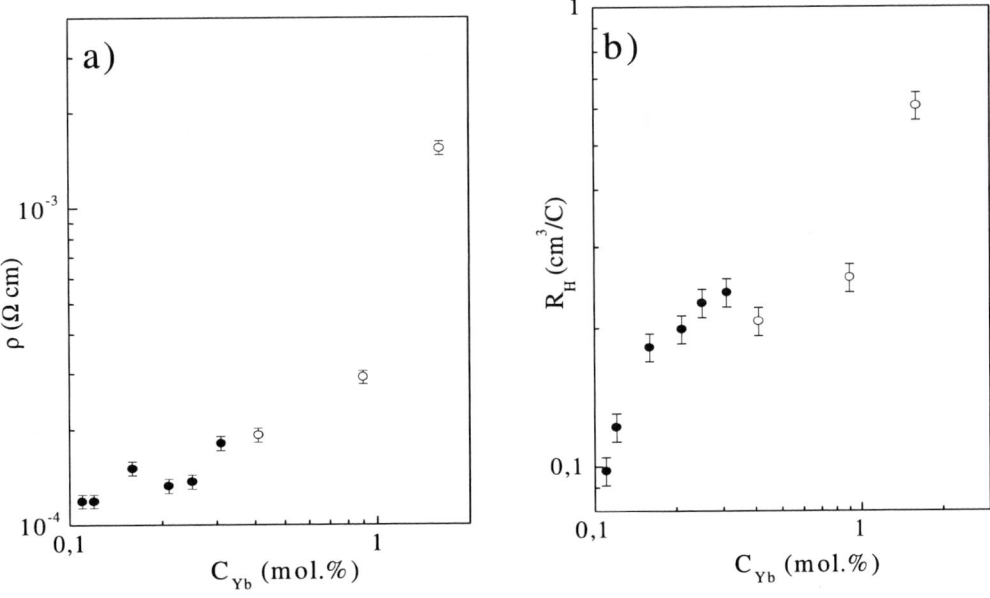

Figure 2. The resistivity (a) and the Hall coefficient (b) versus ytterbium impurity concentration in the $Pb_{1-x}Sn_xTe:Yb$ alloys (black circles – samples 01-15, open circles – samples 17-23).

To explain the data obtained we can assume that like in $Pb_{1-x}Ge_xTe$:Yb alloys, previously investigated [6], doping of $Pb_{1-x}Sn_xTe$ with Yb leads to the appearance of deep defect level in the energy spectrum of the alloys, which stabilizes the FL within the valence band. Dependence of the hole concentration p, determined from the experimental values of the Hall coefficient, on the alloy composition were used to calculate the energy position of the FL E_F relative to the middle of the forbidden band in the frame of two-band dispersion law for A^4B^6 semiconductors [9,10]:

$$\left(\frac{E_g}{2} - E\right) \cdot \left(-\frac{E_g}{2} - E\right) = E_\perp \cdot \frac{p_\perp^2}{2m_0} + E_\| \cdot \frac{p_\|^2}{2m_0} \tag{1}$$

where E_g is the energy gap, m_0 is the free-electron mass, $p_\perp$ and $p_\|$ are transverse and longitudinal components of the quasi-impulse relative to the <111> axis, $E_\perp$ and $E_\|$ are the parameters characterizing an interaction between the valence and conduction bands, which are found for $Pb_{1-x}Sn_xTe$ (x≤0.2) to be equal: $E_\perp \approx 7.65$ eV and $E_\| \approx 0.73$ eV respectively [11]. In this case the concentration of the free holes and the FL may be presented as a follows:

$$p = \frac{4 \cdot 2 \cdot 4\pi p_\perp^2(E_F)p_\|(E_F)}{3 \cdot (2\pi\hbar)^3} \tag{2}$$

$$E_F = \sqrt{\left(\frac{3}{4} \cdot \frac{p\pi^2\hbar^3 E_\perp \sqrt{E_\|}}{(2m_0)^{3/2}}\right)^{2/3} + \frac{E_g^2}{4}} \tag{3}$$

Figure 3 shows the diagram of the FL movement under variation of Yb content in the alloys. It is seen that with increasing Yb concentration the FL non-monotonously approaches the top of the valence band.

Dependencies of the hole concentration and the FL on the ytterbium content allow us to determine now the position of the ytterbium level E_{Yb} in the energy spectrum of alloys under study. It was anticipated that the density of unoccupied states on the ytterbium level p_{Yb}, which is equal to the difference between total amount of the ytterbium introduced N_{Yb} and the electron

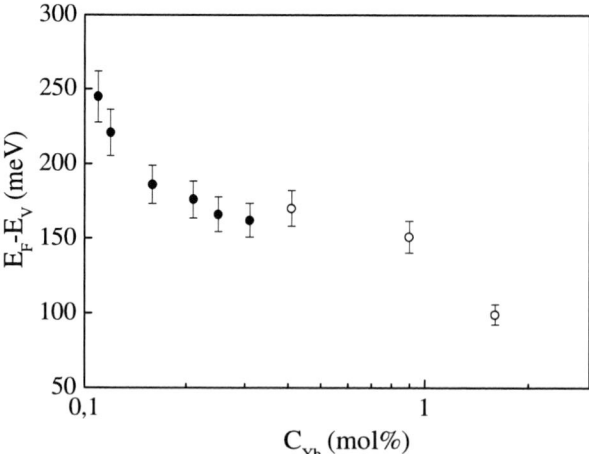

Figure 3. Energy position of the Fermi level in the $Pb_{1-x}Sn_xTe$ alloys at T=4.2 K (black circles – samples 01-15, open circles – samples 17-23).

concentration on the impurity level n_{Yb} coincides with magnetic ions concentration N_{Yb3+} determined earlier [5]:

$$p_{Yb} = N_{Yb} - n_{Yb} = N_{Yb3+} \tag{4}$$

$$p_{Yb} = \int_{E_F}^{\infty} g_{Yb}(E)\,dE \tag{5}$$

where g_{Yb} is the density of states function in the impurity band that may be approximated by a Gaussian-type curve with the impurity bandwidth $\sigma \approx 7$ meV [7]:

$$g_{Yb}(E) = \frac{N_{Yb}}{\sigma\sqrt{2\pi}} \exp\left[\frac{-(E - E_{Yb})^2}{2\sigma^2}\right] \tag{6}$$

Results of our calculation are shown in figure 4. At low ytterbium content (a right section in figure 4) the impurity level is completely filled with electrons and located deeply within the valence band essentially lower the FL. As the density of occupied with electrons states on the Yb level is corresponded to the concentration of the electrically neutral and non-magnetic ions Yb^{2+} it becomes clear why there was observed no paramagnetic response in this case [5]. With increasing C_{Yb} the FL and Yb level move up to the valence band top with different rates (a middle section in figure 4). The impurity level moves faster, intersects the FL and starting at certain critical value of ytterbium concentration stabilizes the FL. Apparently an anomalous leap observed is connected with the intersections of the FL with the impurity band E_{Yb} and pinning of FL by this band. Afterwards the concentration of free states in the impurity band grows due to a flow of electrons from E_{Yb} band to the unoccupied states in the valence band (a left section in figure 4). Therefore the concentration of electrically and magnetically active ions Yb^{3+}, which is equal to the density of empty states on Yb level, increases and we observe a paramagnetic behavior.

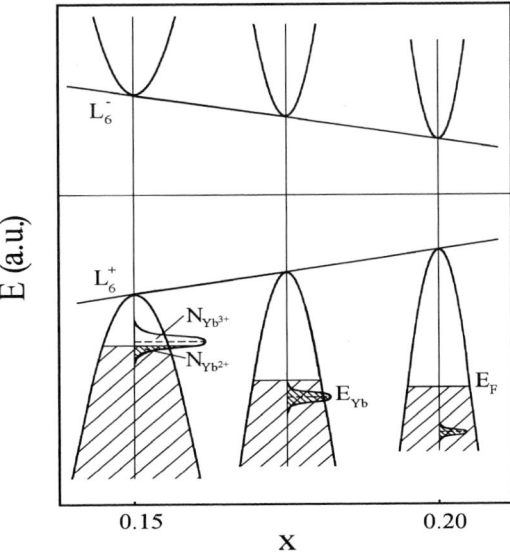

Figure 4. The energy spectrum reconstruction for investigated $Pb_{1-x}Sn_xTe$:Yb alloys upon varying the tin and ytterbium content. Density of states function for ytterbium impurity band was calculated according to the model (4)-(6).

CONCLUSIONS

It was established that doping of the $Pb_{1-x}Sn_xTe$ alloys with ytterbium lead to a formation of deep impurity level (impurity band) E_{Yb}, the position of which depends strongly on the alloy composition. We proposed a model of charge carrier energy spectrum under varying the alloy composition according to which with decreasing of tin content x and increasing ytterbium concentration C_{Yb} the impurity band shifts towards the valence band top, intersects the Fermi level and starting a certain critical value of ytterbium concentration pins it. It was shown that the magnetic properties of $Pb_{1-x}Sn_xTe$:Yb can be successfully explained in terms of the model proposed. These properties depend strongly on the electronic structure of alloys, in particular, on the mutual arrangement of the Fermi level and impurity ytterbium level.

ACKNOWLEDGEMENTS

This research has been supported by the Russian Federation President Program (Grant No SS 1786.2003.2) and Universities of Russia Program (No 01.03.068).

REFERENCES

1. B.A. Volkov, L.I. Ryabova, D.R. Khokhlov, *Physics-Uspekhi* **45**, 819 (2002).
2. Lead Chalcogenides: Physics and Applications ed. by D. R. Khokhlov in *Optoelectronic properties of semiconductors and superlattices* **18** (Taylor and Francies, New York, London, 2003).
3. T. Story, *Acta Phys. Pol. A* **94**, 189 (1998).
4. V.K. Dugaev, V.I. Litvinov, A. Lusakowski, *Phys. Rev. B* **59**, 15190 (1999).
5. E.P. Skipetrov, E.A. Zvereva, L.A. Skipetrova, B.B. Kovalev, O.S. Volkova, A.V. Golubev, and E.I. Slyn'ko, *phys. stat. sol. (b)* **241**, 1100 (2004).
6. E.P. Skipetrov, N.A. Chernova and E.I. Slyn'ko, *Phys. Rev. B* **66**, 085204 (2002).
7. E.P. Skipetrov, N.A. Chernova, E.I. Slyn'ko and Yu.K. Vygranenko, *Phys. Rev. B* **59**, 12928 (1999).
8. V.E. Slyn'ko, Visnyk Lviv Univ., *Ser. Physic.* **34**, 291 (2001).
9. R. Dornhaus, G. Nimtz and B. Schlicht, *Narrow-Gap Semiconductors* (Springer-Verlag, Berlin, 1983) pp.45-49.
10. E. O. Kane, *J. Phys. Chem. Sol.* **1**, 249 (1957).
11. B.A. Akimov, R.S. Vadhva, S.M. Chudinov, *Sov.-Phys. Semicond.* **12**, 1927 (1978).

Mat. Res. Soc. Symp. Proc. Vol. 825E © 2004 Materials Research Society

Magnetic Properties of GaN Layers Implanted by Mn, Cr or V.

Vitaliy A. Guzenko[1], Nicolas Thillosen[1], Andre Dahmen[1], Raffaella Calarco[1], Thomas Schäpers[1], Martina Luysberg[2], and Lothar Houben[2]

[1]Institute of Thin Films and Interfaces (ISG-1), Research Centre Jülich, 52425 Jülich, Germany
[2]Institute of Solid State Research (IFF), Research Centre Jülich, 52425 Jülich, Germany

ABSTRACT

We report on magnetic properties of the GaN layers implanted with 3d transition metal ions. GaN layers grown by MOVPE on sapphire substrates, p- or n-doped, were implanted by Mn, Cr or V ions with a dose of 5×10^{16} cm^{-2} and implantation energy of 200 keV. Subsequently, a rapid thermal annealing in nitrogen atmosphere for 5 minutes at different temperatures (700°C – 1050°C) was performed. The magnetization as a function of magnetic field as well as the dependence on temperature revealed paramagnetic behavior for all samples. In addition, an antiferromagnetic coupling between implanted ions was found.

INTRODUCTION

In recent years the field of spintronics has been rapidly developing and a special role has been played by diluted magnetic semiconductors (DMS), owing to their potential for future spintronic devices [1]. In comparison with conventional ferromagnets they are of advantage for an efficient electrical spin injection into semiconductors, because of the better material matching. Recently, GaN-based DMS became of great interest due to prediction of room temperature ferromagnetism in GaN:Mn by Dietl et al. [2] and other compounds (GaN:Cr and GaN:V) by Sato et al. [3]. It has been proposed that a stable ferromagnetic state in GaN:Mn (Mn concentration up to 10%), in GaN:Cr and GaN:V (at even higher impurity concentration) should be achieved. Interestingly, according to the band structure calculations by Sato et al. [3], ferromagnetic ordering, which pertains to the partial filling of the anti-bonding impurity states, in n-type GaN:V seems to be possible.

Several groups have investigated the growth and properties of GaN-based DMS. However, the results show significant discrepancies, particularly regarding the magnetic properties of the layers. It has been reported on ferromagnetic [4–8] or paramagnetic [9], antiferromagnetic [10] or spin-glass [11] behavior. It is worth to be mentioned that different techniques were applied for the fabrication of GaN-based DMS: molecular beam epitaxy [9, 11], metal-organic chemical vapor deposition [4, 5], ammono-thermal technique [10], or ion implantation [6–8]. The origin of the ferromagnetism in GaN-based DMS is still far from being completely understood. Hence, the question, whether it is possible to fabricate homogeneous ferromagnetic layers without precipitation, is of particular importance.

Between others one of the basic prerequisites for the realization of spintronic devices is the electrical injection of the spin-polarized current into GaN. To meet this criterion good ohmic contacts are necessary. Because of the wide band gap and strong Fermi level pinning a large Schottky barrier [12] in the case of p-type GaN is present, whereas a preparation of the ohmic contacts to n-type GaN is well developed. Thus, in spite of the fact that vanadium ions possess a magnetic moment smaller then that of manganese, n-type GaN:V might be of advantage for the preparation of ohmic contacts and electrical spin injection.

These main motivations pushed us to study the magnetic properties of n-type GaN layers implanted with Mn, Cr, or V ions in comparison with p-type GaN implanted with Mn. In addition to the measurements of the magnetic properties, the crystalline structure was investigated by means of transmission electron microscopy (TEM).

RESULTS AND DISCUSSION

The GaN layers were grown by MOCVD on a (0001) sapphire substrate. Two types of GaN films were used: a 1.5 μm thick n-type GaN layer, doped with Si and a 1.0 μm thick p-type GaN layer, doped with Mg. In both cases the doping concentration was 2×10^{17} cm^{-3}. The n- and p-doped GaN layers are separated from the sapphire substrate by a 3 and 1 μm thick GaN buffer layer, respectively. Mn^{+}, Cr^{+}, or V^{+} ions were implanted with an energy of 200 keV and a dose of 5×10^{16} cm^{-2}. During the implantation the samples were held at a temperature of 350°C, in order to prevent amorphization. Subsequently, a rapid thermal annealing for 5 min in N$_2$ atmosphere at temperatures in the range of 700°–1050°C was performed. To remove residual unintentional ferromagnetic contaminations before the annealing step all the samples were cleaned in hydrochloric acid.

The magnetization of the samples was measured by a SQUID magnetometer (Quantum Design MPMS7) operating in the temperature range of 1.7 – 400 K and magnetic fields up to 7 T, applied parallel to the sample surface. Particular attention has been paid to enhance the sensitivity of the experimental setup and to reduce the systematic error due to background signal stemming from the sample holder. For this purpose a quartz glass rod with a diameter of 0.8 mm was designed. The rod was cleaned with hydrofluoric acid to remove eventual ferromagnetic contaminations. The samples were fixed by a drop of glue, which was checked to be diamagnetic and was placed in the middle of the sample to reduce distortions of the measured SQUID signals.

The magnetization curves for Mn implanted n-type and p-type GaN samples, measured at the temperatures of 2 K, 10 K and 100 K, are displayed in figure 1. The data show a typical paramagnetic behavior and can be fitted well by the Brillouin function for temperatures of 3 K, 10 K and 100 K, assuming a spin S=5/2. The diamagnetic background signal was subtracted by measuring the magnetization curve of the corresponding as-grown samples. Between the initially p- and n-doped samples no considerable difference could be observed.

A reasonable fit of the experimental data for the lowest temperature T_{exp}= 2 K has been achieved only if a slightly higher effective temperature T_{eff} of 3 K was assumed. At higher temperatures the difference between T_{eff} and T_{exp} is negligible. Moreover, the saturation magnetization is different from the calculated value. The experimentally determined magnetic moment at 2 K amounts to be smaller by a factor of 0.22 compared to the theoretically expected one. In the following we denote this factor as x_{eff}.

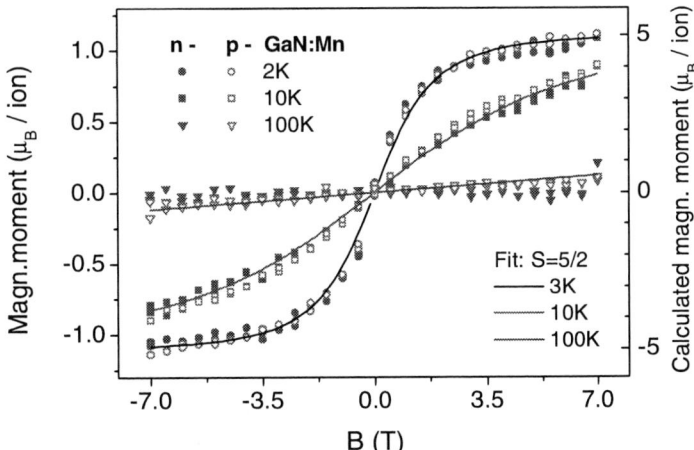

Figure 1. Magnetization curves of the p-type and n-type GaN implanted with Mn at 2 K, 10 K, and 100 K. The right y-axis represents calculated values of the saturation magnetization.

A similar effect as described above was observed by Graf *et al.* [9] and Zając *et al.* [10] in GaN:Mn samples with different Mn content. For highly diluted GaN:Mn, where the spatial separation of spins is large, the magnetization data could be described by the classical Brillouin function $B_S(x)$. In samples with a high 3d metal impurity content the magnetic ions are found closer to each other and can consequently interact stronger. Hence, pairs and/or larger clusters of antiferromagnetically ordered spins are formed. For these samples an effective temperature T_{eff} and an effective Mn content Nx_{eff} with $x_{eff} < 1$ can be introduced, so that the magnetization can be expressed by:

$$M = Nx_{eff}\, g\mu_B S B_S\left(B, T_{eff}\right),$$

where N is the total number of the implanted ions, $g \approx 2$ is the Landé factor and μ_B is the Bohr magneton. The antiferromagnetically coupled spins respond to the applied magnetic field less effectively then the non-interacting, leading to an effective temperature T_{eff} larger than the experimental temperature T_{exp}. Moreover, the total spin of such clusters is zero or equal to the spin of the single ions, depending on the number of ions in a particular cluster. As a result, a lower saturation magnetization is measured, which can be expressed by lower concentration ($x_{eff} < 1$).

A comparison of the magnetization curves of the n-type GaN samples implanted with Mn, Cr, and V, measured at 2 K and 10 K, reveals similar behavior for the GaN:Cr (S=2) and GaN:V (S=3/2) samples (see figure 2). The measured saturation magnetization of these samples is lower than of the Mn implanted sample. However, the ratio of the experimentally determined and calculated values of the saturation magnetization is slightly higher. The highest ratio $x_{eff} = 0.28$ at 2 K shows the V implanted GaN sample. The corresponding value for the Cr implanted sample is $x_{eff} = 0.23$. It is remarkable that the value of x_{eff} reduces with an increasing ion spin. We attribute this to a stronger interaction between ions with a larger spin. Since the implanted ions couple antiferromagnetically, their response to the applied magnetic field is the highest for the case of GaN:V and the lowest for GaN:Mn. This is then reflected in the value of x_{eff}. At higher measurement temperatures the antiferromagnetic coupling between 3d ions grows weaker, which results in an increase of x_{eff}.

The temperature dependencies of the magnetic moment of all GaN:X (X = Mn, Cr or V) samples measured at a constant magnetic field of 50 mT show also a paramagnetic behavior. The experimental curves are shown in figure 3. All data can be fitted well by the Curie-Weiss law $1/(T-\Theta)$. The corresponding Curie-Weiss temperatures Θ are negative, which is an indication of the antiferromagnetic coupling between implanted ions.

It is noticeable that all samples both initially p- and n-doped became extremely highly resistive or even insulating after ion implantation. This can be attributed to deep electron traps, occurring due to the formation of complexes between the transition metal ions and some native defects as well as due to radiation damage defects not involving transition metal ions [13].

The theoretical predictions of the ferromagnetic ordering in GaN-based DMS rely on the assumption that a high density of valence band holes is provided; otherwise the antiferromagnetic or spin glass state is stable. Since all samples prepared in this study are highly resistive, i.e. the concentration of the free carriers is low, an antiferromagnetic, rather than a ferromagnetic coupling between implanted ions is expected. Furthermore, since the concentration of 3d ions exceeds the solubility limit in GaN the formation of an antiferromagnetic secondary phase during the annealing procedure cannot be excluded. The presence of precipitates in the top layer was demonstrated by high-resolution TEM, not shown here [14]. However, detailed study of the crystalline structure as well as the chemical composition is necessary to draw conclusions concerning their magnetic properties.

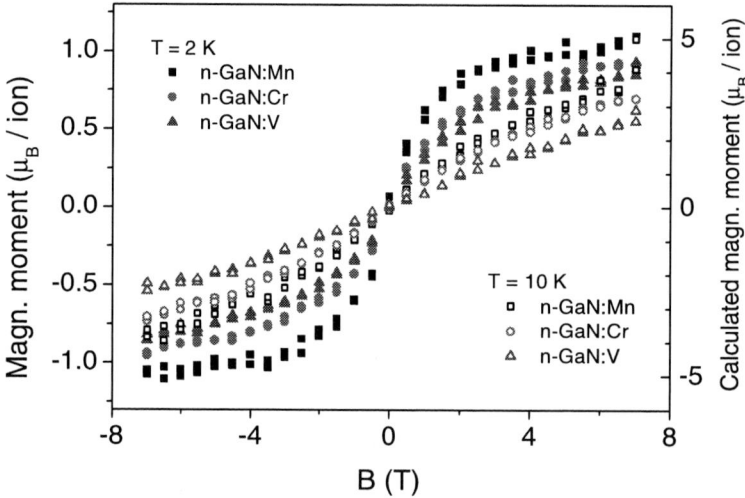

Figure 2. Magnetization curves of the n-type GaN implanted with Mn, Cr, or V at 2 K and 10 K. The right y-axis represents calculated value of the saturation magnetization.

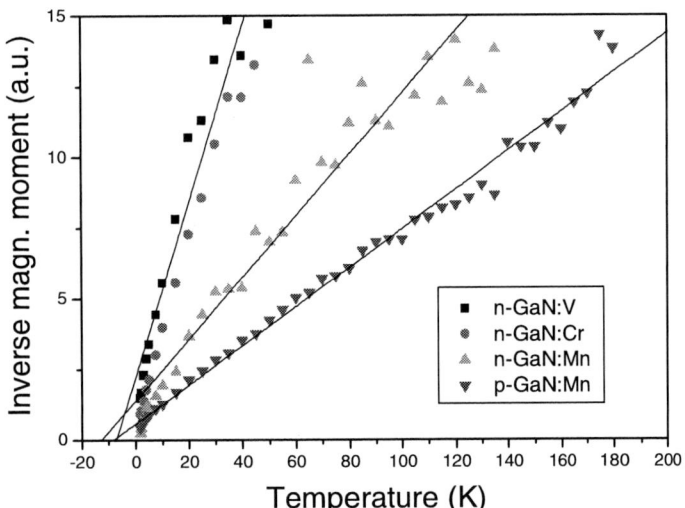

Figure 3. Temperature dependencies of inverse magnetic moment of the GaN:X (X = Mn, Cr, V) layers in a constant magnetic field of 50 mT.

CONCLUSIONS

We have prepared GaN layers implanted with Mn, Cr, or V by means of ion implantation and subsequent annealing for 5 min in a flowing N_2 atmosphere in the range of temperatures between 700°C and 1050°C. The studies of magnetic properties of these samples reveal paramagnetic behavior with an antiferromagnetic coupling between the 3d metal ions. This behavior is owing to the low concentration of the delocalized carriers necessary to mediate the ferromagnetism and formation of the clusters, which interacts antiferromagnetically. No indications of a ferromagnetic ordering of the implanted ions could be found.

REFERENCES

1. H. Ohno, *Science* **281**, 951 (1998).
2. T. Dietl, H. Ohno, F. Matsukura, J. Cibert, and D. Ferrand, *Science* **287**, 1019 (2000).
3. K. Sato and H. Katayama-Yoshida, *Jap. J. Appl. Phys.* **40**, L485 (2001).
4. M. L. Reed, N. A. El-Masry, H. H. Stadelmaier, M. K. Ritums, M. J. Reed, C. A. Parker, J. C. Roberts, and S. M. Bedair, *Appl. Phys. Lett.* **79**, 3473 (2001).
5. M. L. Reed, M. K. Ritums, H. H. Stadelmaier, M. J. Reed, C. A. Parker, S. M. Bedair, and N. A. El-Masry, *Mat. Lett.* **51**, 500 (2001).
6. N. Theodoropoulou, A. F. Hebard, M. E. Overberg, C. R. Abernathy, S. J. Pearton, S. N. G. Chu, and R. G. Wilson, *Appl. Phys. Lett.* **78**, 3475 (2001).
7. J. S. Lee, J. D. Lim, Z. G. Khim, Y. D. Park, S. J. Pearton, and S. N. G. Chu, *J. Appl. Phys.* **93**, 4512 (2003).

8. J. M. Baik, J. K. Kim, H. W. Jang, Y. Shon, T. W. Kang, and J.-L. Lee, *Phys. Stat. Sol. B* **234**, 943 (2002).

9. T. Graf, M. Gjukic, M. Hermann, M. S. Brandt, M. Stutzmann, L. Görgens, J. B. Philipp, and O. Ambacher, *J. Appl. Phys.* **93**, 9697 (2003).

10. M. Zając, J. Gosk, M. Kaminska, A. Twardowski, T. Szyszko, and S. Podsiadlo, *Appl. Phys. Lett.* **79**, 2432 (2001).

11. S. Dhar, O. Brandt, A. Trampert, L. Daweritz, K. J. Friedland, K. H. Ploog, J. Keller, B. Beschoten, and G. Güntherodt, *Appl. Phys. Lett.* **82**, 2077 (2003).

12. J. Rennie, M. Onomura, S.-Y. Nunoue, G.-I. Hatakoshi, H. Sugawara, and M. Ishikawa, *J. Cryst. Growth* **189-190**, 711 (1998).

13. A. Y. Polyakov, N. B. Smirnov, A. V. Govorkov, N. Y. Pashkova, A. A. Shlensky, S. J. Pearton, M. E. Overberg, C. R. Abernathy, J. M. Zavada, and R. G. Wilson, *J. Appl. Phys.* **93**, 5388 (2003).

14. V.A. Guzenko, N. Thillosen, A. Dahmen, R. Calarco, Th. Schäpers, L. Houben and M. Luysberg, A. Kaluza, and B. Schineller (to be published).

Mat. Res. Soc. Symp. Proc. Vol. 825E © 2004 Materials Research Society G5.5.1

Optimization of Sample Holder Materials for Sensitive Magnetometry Measurements at Low Temperatures

I. Bossi, N.R. Dilley, J. R. O'Brien, and S. Spagna.
Quantum Design Inc.,
6325 Lusk Blvd., San Diego, CA, 92121

ABSTRACT

Magnetization measurements were performed as a function of magnetic field H and temperature T on samples of nine different materials including clear fused quartz, cartridge brass, G-10 glass-reinforced epoxy, acetal homopolymer, glass-filled acetal, phenolic, and other plastics. A small yet distinct amount of ferromagnetic or paramagnetic impurities is observed in all the materials investigated in this study except quartz. In contrast, the magnetic response of quartz is typical of a diamagnet over the temperature range 5 K to 300 K. The volume susceptibility is equal to -4.4×10^{-7} (cgs) over the whole temperature range.

INTRODUCTION

Vibrating sample magnetometers (VSM) are fast and sensitive instruments, commonly used to characterize magnetic materials in terms of their hysteresis loop M(H), and magnetization versus temperature M(T) measurements. VSM systems find a broad application in several fields, including magnetic recording research [1], superconductivity [2, 3], and nanotechnology [4]. In general, when measuring the d.c. magnetic moment of a sample using a VSM, the total moment measured is:

$$m = m_s + m_{sh} + m_e \qquad (1)$$

where m_s is the magnetic moment of the sample, m_{sh} the magnetic signal from the sample holder, and m_e the magnetic signal from the environment, such as parasitic magnetic signals due to synchronous vibrations of the equipment. A crucial prerequisite for the accuracy of the measurement is that $m_s >> m_{sh} + m_e$. To our knowledge, little published work has been dedicated to the systematic investigation and the optimization of state of the art engineered sample holder materials for purpose of performing sensitive magnetometry measurements.

The objective of this study is to characterize the magnetic behavior of materials in an effort to find those which produce a low magnetic signal m_{sh} when used as sample holders for VSM measurements. The requirements for VSM sample holder materials, other than having a low concentration of ferromagnetic or paramagnetic impurities, include cleanliness, large temperature range of operation (usually 1.9 K to 400 K), manufacturability (for example the ability to produce a rigid and straight sample holder that will not rub inside the detection coil set), low cost, longevity, and ease of sample mounting.

EXPERIMENTAL DETAILS

This paper focuses on the characterization of the following nine different materials: (1) clear fused quartz rod [5], (2) gold-plated cartridge brass tubing [6], (3) G-10 glass cloth-reinforced epoxy (G-10) rod, (4) graphite-and PTFE-filled polyetheretherketone (PEEK) rod, (5) acetal homopolymer rod, (6) cotton cloth reinforced phenolic plastic rod, (7) bearing-grade polyphenylene (PPS) rod [7], (8) G-10 tubing, and (9) glass-filled acetal rod.

In this investigation, both rod and tube starting materials geometries were considered as they both are useful in fabricating sample holders for a VSM (see figure 1). Of the rod-shaped starting materials, clear fused quartz can be used to fabricate a VSM sample holder by grinding a rod to a paddle-shaped holder to produce a flat surface for sample adhesion [5]. Gold-plated cartridge brass and G-10 tubing can be machined along the length to produce a trough with a cross section shaped like a "C" that can be used to clamp a cylindrical sample.

Figure 1. Sketch of the two sample holder geometries evaluated in this study.

Small specimens of the nine materials under investigation were obtained in the shape of cylinders or thin rectangular platelets with a typical volume between 4 and 170 mm^3. After ultrasonic cleaning in alcohol for a few minutes, the samples were mounted into a clear straw or attached to a thin quartz rod with kapton tape. The magnetic moment of the sample holder for this experiment was measured to ensure perfect diamagnetic behavior, uniformity and negligible magnitude compared to the sample magnetic signal.

SQUID magnetometer measurements

Isothermal magnetization curves M(H) at 5 K and 300 K were obtained using a 5 tesla Magnetic Properties Measurement System (MPMS). In this system, a superconducting (SC) magnet, sitting in a liquid helium bath, generates a uniform longitudinal applied magnetic field. The magnetic flux signal from a moving sample is detected using a SQUID based detection system, comprising of a SQUID sensor coupled to SC pick-up loops through a SC transformer. The sample is moved through the center of the pick-up loops, which are arranged in a second derivative configuration to extract the absolute value of the sample moment. In the magnetometer, cold helium gas or liquid is drawn from the cryostat Dewer through a cooling annulus surrounding the sample environment. The gas is pre-heated to the desired sample temperature and it is flowed upwards to provide a uniform temperature control. In general, the sample temperature can be controlled in this way in the temperature range between 1.9 K and 400 K. Upon centering the sample within the SC pick-up loops at room temperature in a field of

2 tesla, Reciprocating Sample Option [8] magnetic moment measurements were performed while the sample was oscillated at 1.5 Hz over a scan length of 4 cm. Using a combination of auto tracking and iterative regression in the analysis of the output data the system compensates for the thermal contraction of the sample rod upon cooling the sample to 5 K. This capability of the magnetometer allowed collecting data at all temperatures, with no need of re-centering the sample within the SC pick-up loops. The magnetic field was varied from –5 tesla to +5 tesla, in hysteresis mode in order to rapidly set the applied magnetic fields.

VSM measurements

Isothermal magnetization curves M(H) at T = 5 K and 300 K were obtained using the VSM option for the Physical Properties Measurement System (PPMS). While the sample was oscillated at 40 Hz and 2mm peak amplitude, the magnetic field was swept continuously from –5 tesla to +5 tesla.

DISCUSSION

Table I compares the experimental value of the volume susceptibility of the nine materials investigated in this study. Magnetization versus applied magnetic field data collected at T = 5 K and T= 300 K for gold-plated brass [6], and clear fused quartz are compared in Figure 2a and 2b respectively. The presence of ferromagnetic impurities in brass is noticeable, especially at T = 5 K. In contrast, from the data in Figure 2b, it is apparent that quartz acts as a perfect diamagnet both at room and low temperature. From Table I, the volume susceptibility of the brass at low temperature is about one order of magnitude larger than that at room temperature. At T = 5 K as the applied field is increased, the induced magnetization increases and saturates around 5 tesla ($M_s \sim 0.22$ emu/cm^3). As the applied field is reversed, the induced magnetization decreases but is different than zero when the applied field is removed. The remnant magnetization is $M_r = 1.55 \times 10^{-2}$ emu/cm^3 at 5 K. As the applied field is further decreased, the magnetization is completely removed at H_c = -200 Oe. The squareness (SQ=M_r/M_s) is equal to SQ=0.07. At room temperature a small ferromagnetic component becomes visible. As the field is increased, the magnetization increases until 2000 Oe, when the diamagnetic component prevails. Continuing increasing the field, the magnetization of the sample has two components, diamagnetic and ferromagnetic. However, the latter component is saturated and has no influence on the slope of the magnetization characteristics at fields over

Table I. Volume susceptibility χ(cgs) = M(emu/cm^3)/H(Oe) at high and low temperature

Material	M/H at 300 K	M/H at 5 K
quartz rod	-4.35×10^{-7}	-4.31×10^{-7}
gold-plated brass tubing	4.97×10^{-6}	2.71×10^{-5}
G-10 rod	1.52×10^{-7}	4.25×10^{-5}
PEEK rod	2.75×10^{-5}	2.9×10^{-5}
acetal rod	-4.07×10^{-7}	6.45×10^{-6}
phenolic rod	-6.35×10^{-7}	-2.51×10^{-7}
PPS rod	-1.57×10^{-7}	6.09×10^{-6}
G-10 tubing	-1.20×10^{-6}	2.15×10^{-5}
Glass-filled acetal rod	-4.22×10^{-7}	6.69×10^{-6}

2000 Oe. When the applied field is reversed, the magnetization first decreases linearly, then below 2000 Oe the ferromagnetic component dominates, with $M_r=1.55\times10^{-3}$ emu/cm^3 at H=0. The coercitivity is $H_c\sim-80$ Oe.

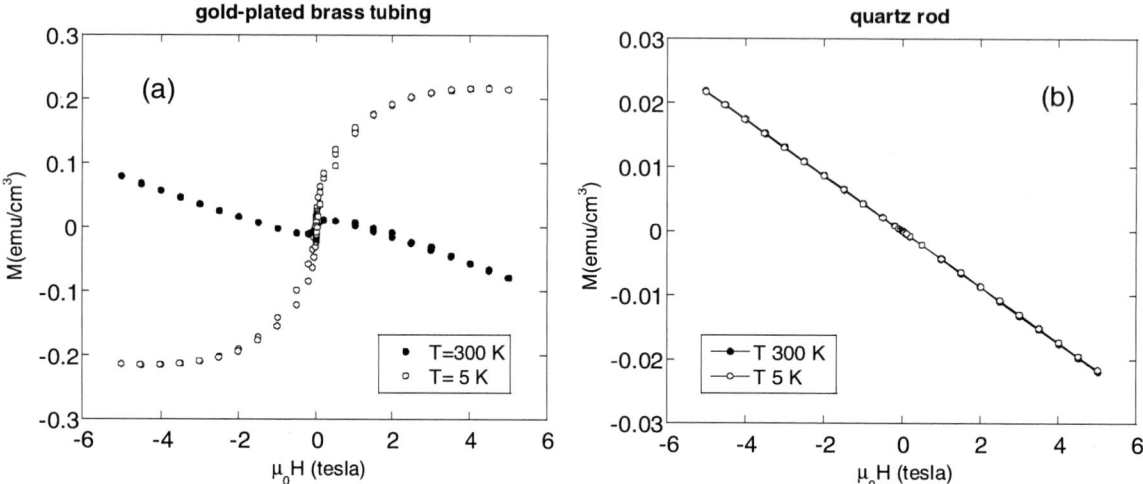

Figure 2. **(a)** Magnetization of a small gold-plated brass tubing specimen as a function of applied magnetic field at 300 K and 5 K showing the presence of ferromagnetic and paramagnetic impurities which becomes evident at low temperature. **(b)** Magnetization of a small quartz rod specimen as a function of applied magnetic field at 300 K and 5 K, revealing perfect diamagnetic behavior at both temperatures. All the above data are obtained using the MPMS SQUID magnetometer.

Traces of ferromagnetic impurities are visible in the G-10 rod at T=300 K and low fields (H<4000 Oe). At T=5 K the magnetization is dominated by a paramagnetic term which is large in magnitude as compared to the ferromagnetic signal.

The graphite-and PTFE-filled PEEK plastic showed the presence of ferromagnetic impurities at both high and low temperature, with $M_r = 3.03\times10^{-3}$ emu/cm^3, $M_s = 6.2\times10^{-2}$ emu/cm^3, $H_c=-150$ Oe and SQ=0.05 at T=300 K. These parameters at T =5 K are $M_s = 8.5\times10^{-2}$ emu/cm^3, $M_r = 8.4\times10^{-3}$ emu/cm^3, Hc = -330 Oe and SQ = 0.098.

Acetal, PPS, G-10 tubing, and glass filled acetal at T = 300 K are predominantly diamagnetic, while at low temperature these materials have a large paramagnetic component as we can see from Table I. Phenolic is primarily diamagnetic, with a small paramagnetic component showing at T=5 K.

When these materials are used to make sample holders for the VSM, one requirement is to make them long (100mm) with respect to the distance between the detection coils (9mm). A sample holder of uniform cross-section and with a uniform distribution of magnetic impurities will thus appear more transparent to the VSM detection than a small, localized specimen. VSM sample holders were made from both quartz rod and brass tubing and were magnetically characterized at room temperature and low temperature using the VSM. The sample holder was vertically centered in the coilset such that the end of the sample holder was sufficiently far from and $\Psi_{lh^-}^m(z, \rho, N_L)$. The first occurrence of states hh_0^+ is in the multiplet with $N_L = 2$, and has interaction with its partners lh_1^+, lh_2^- and hh_3^- for the different values of m. The dispersions for this multiplet can show anti-crossings regions at specific values of magnetic field, a result directly linked to their coupling induced by the form of the Luttinger-Kohn Hamiltonian. Finally, we should point out that there will be no coupling between states belonging to different multiplets and, thus, they may show crossing (degeneracy) for any B_0.

The Landau-Stark dispersions, for a 50 nm GaAs/Al$_{0.35}$Ga$_{0.65}$As isolated QW, calculated with the model just described are shown in Figs. 1 and 2 for small ($F_0 = 10$ kV/cm) and large ($F_0 = 100$ kV/cm) electric field, respectively. The corresponding Landau numbers and spin polarizations are indicated in the figure. Observe the combined Zeeman and Stark effects on these mixed levels where the time-reversal symmetry is broken by both fields, leading to anomalous dependence of all $g_{hh_{n_L}(lh_{n_L})}(B_0, F_0)$ with applied fields. This happens once the electric field pushes the eigenvalues to the bottom of the QW and thus, increases the coupling between levels belonging to each multiplet discussed above. The shape of anticrossing regions at specific values of B_0 as well as the crossing between levels in different N_L multiplets can be observed in these figures.

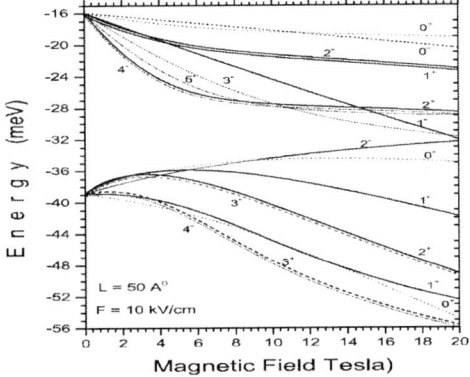

Landau-Stark magneto-dispersions for the first heavy- and ligth-hole states, at $F_0 = 10$ kV/cm. Notice the level crossing, anticrossing and anomalous g-factors for different multiplets discussed in text.

Same as in Fig.3, with larger field F_0. Observe changes in Zeeman separtion induced by electric compression of states.

Furthermore, there is a change of sign in g_{hh_0} for magnetic fields above 12 T and where the Zeeman splitting becomes clearly resolved. The same effect occurs for g_{hh_1} above 4 T and for g_{lh_2} above 3 T. In general, the light-hole Landau-Stark states have a larger Zeeman splittings than the corresponding heavy-hole multiplet for any given magnetic field.

Stronger electric fields cause many changes and new distortions to the energy dispersions of all states. First, the "level packing" effect increases the interaction between states and leads to the change of sign and nonlinear values for the g-factors. At this large electric field, there is complete resolution of the hh_0^- and hh_0^+ Zeeman splitting as well as the suppression of the sign change for this Kramer doublet. However, the doublet hh_1^-, hh_1^+ (hh_2^-, hh_2^+) displays two (three) regions where is observed sign change in the g-factor, an effect purely caused by "level compression" induced by electric field.

As the magnetic field increases, we observe a spreading on the dispersion curves with the lh branches occurring in a wider interval of the energy as compared to the hh branches, since they will be magnetically "compressed" by some lh states belonging to the same interacting multiplet. Above 15 T the energy of the lh_0^- states is found below the energies of the heavy-hole states with higher Landau number, hh_4^-, hh_5^-. Since they are in different sets, they can show crossing regions at any magnetic field.

A. Magneto-current

Having shown the complexity of the 2D Landau-Stark dispersions let us now explore these spin-splittings on the vertical transport of holes in DBRST structure with interfaces [thicknesses] at positions z_ℓ ($\ell = 1, ..., 4$) $[d_l = (z_{\ell+1} - z_\ell)]$

the center of the coils (35mm) that the end effects from the sample holder were minimized. This location (35mm) is also typically where a sample would be placed on the sample holder. Figure 3 below reveals the interesting fact that although the quartz material appears to be magnetically much cleaner than gold-plated brass when inspected as small specimens (see Table I), the signal for the two is very similar when they are measured in the VSM as sample holders. One advantage of the brass, owing to its physical strength and springiness, is that it can be made in thin-walled tubing and thus involve a very small volume of material.

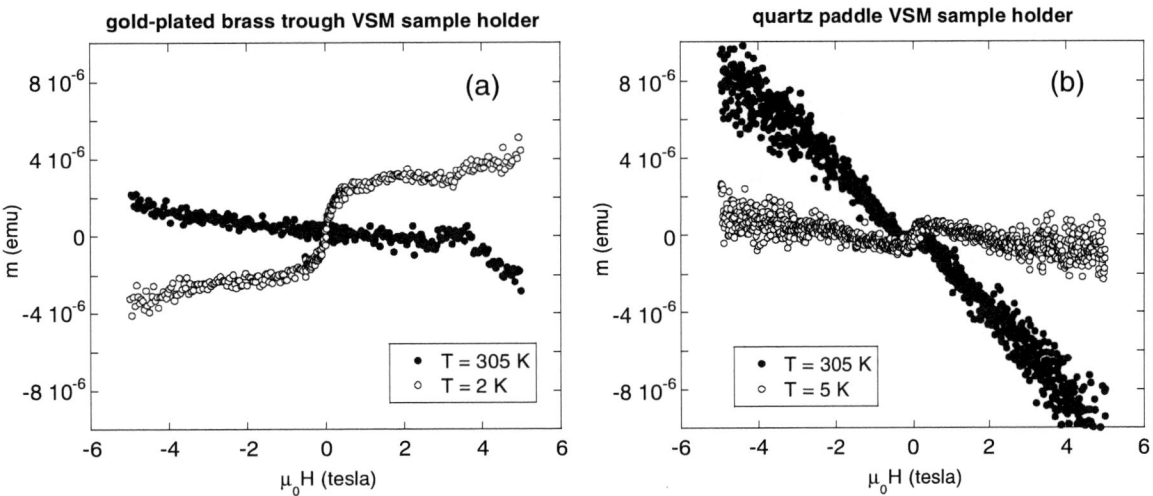

Figure 3. (**a**) Magnetization of a gold-plated brass trough VSM sample holder as a function of applied magnetic field at room temperature and low temperature. (**b**) Magnetization of a quartz paddle VSM sample holder as a function of applied magnetic field at room temperature and low temperature. In both samples a small saturable paramagnetic impurity is visible at low temperature, which is slightly higher in brass ($M_s = 2 \times 10^{-6}$ emu) than in quartz ($M_s = 1 \times 10^{-6}$ emu). Measurements are performed while sweeping the field from -5 tesla to $+5$ tesla at 200 Oe/sec (a, left panel) or 100 Oe/sec (b, right panel). Data collected while sweeping from $+5$ tesla to -5 tesla (not shown) do not indicate any measurable magnetic hysteresis for either sample holder. Data are obtained using the PPMS VSM.

CONCLUSIONS

This study showed that quartz exhibits optimal magnetic behavior, which renders it one of the most favorable materials for manufacturing VSM sample holders. On the other hand gold-plated brass tubing and G-10 tubing or rod are more robust, cheaper and easier to machine than quartz. These three materials have been used to fabricate the sample holders for VSM option for the PPMS. Once the material is machined into a sample holder much longer compared to the distance between the VSM detection coils, the effect of the impurities will be negligible as far as they are uniformly distributed. The other plastics weren't used primarily because they are not rigid enough to produce a holder, which needs to oscillate at 40 Hz in the 6.3 mm bore VSM detection coil set without rubbing.

REFERENCES

1. N.C. Tansil, R.V. Ramanujan, H.F. Li, Transactions of the Indian Institute of Metals **56**, 509-512 (2003).
2. T.A. Prikhna, W. Gawalek, V.E. Moshchil, L.S. Uspenskaya, R. Viznichenko N.V. Sergienko, A.A. Kordyuk, V.B. Sverdun, A.B. Surzhenko, D. Litzkendorf, T. Habisreuther, A.V. Vlasenko, Physica C **392**, 432-436 (2003).
3. I. Bossi, M.S. Thesis, San Diego State University, 2003.
4. X.Y. Yuan, G.S. Wu, T. Xie, Y. Lin, L.D. Zhang, Nanotechnology **15**, 59-61 (2004).
5. Type 214 clear fused quartz from General Electric. Rod was ground lengthwise to produce a flat paddle surface by GM Associates, Inc., Oakland, CA 94603.
6. Alloy CDA260 from K&S Engineering, Chicago, IL 60638. Gold-plating was performed using a nickel-free gold strike with a cobalt-hardened Orosene 999 bath.
7. Techtron HPV from Quadrant Engineered Plastics, Reading, PA 19612.
8. J. Diederichs, S. Spagna, and R. E. Sager, Czech. J. Phys. **46 S5**, 2803 (1996).

Rashba spin-orbit coupling in InGaAs/InP quantum wires

Jens Knobbe, Vitaliy A. Guzenko and Thomas Schäpers
Institute of Thin Films and Interfaces (ISG-1), Research Centre Jülich,
52425 Jülich, Germany

ABSTRACT

The effect of Rashba spin-orbit coupling on the transport properties of InGaAs/InP quantum wire structures is investigated. The geometry of the wire structures was defined by selective wet chemical etching. For wires without a gate a clear beating pattern, due to the presence of the Rashba spin-orbit coupling, is observed for wires with a width down to 600 nm. For narrower wires no beating pattern is found. The experimental observations are explained by contribution of the Rashba spin-orbit coupling to the one-dimensional magnetosubbands. By depleting the one-dimensional conductor by means of a gate electrode the Rashba coupling strength could be controlled.

INTRODUCTION

Spin related phenomena in semiconductors have attracted a lot of attention in recent years, owing to their potential for future spin electronic circuits [1]. For a successful implementation of spin electronic devices two basic requirements have to be fulfilled. First, it must be possible to inject spins into the semiconductor structure, which are well aligned along a particular orientation. This can be achieve by means of metallic ferromagnetic electrodes or by dilute magnetic semiconductors [2]. The second requirement is an efficient control of the electron spin within the device. In this respect the Rashba effect [3] is an interesting option, since it allows to adjust the spin precession of electrons in a two-dimensional electron gas by applying a gate voltage. An often cited example of a spin electronic device based on the Rashba effect is the spin transistor proposed by Datta and Das [4]. Following this approach, many novel device concepts have been developed [5-7].

The Rashba effect originates from the spin-orbit coupling due to a macroscopic electric field in a two-dimensional electron gas (2DEG). The Rashba spin-orbit coupling leads to a spin-splitting of the electron subbands in the 2DEG according to:

$$E_z = E_z^{sub} + \frac{\hbar^2 k^2}{2m^*} \pm \alpha_R |k|, \tag{1}$$

where E_{sub} denotes the subband energy of the 2DEG, m^* is the effective electron mass and k is the wave vector within the plane of the 2DEG. The strength of the Rashba spin-orbit coupling is quantified by the parameter α_R. The Rashba effect is most pronounced in low band gap channel layers like InAs or InGaAs. According to Eq. (1) the spin-splitting leads to a characteristic beating pattern in the Shubnikov-de Haas oscillations of the 2DEG, due to the splitting into two subbands with different effective electron concentrations [8,9]. It was demonstrated that the spin-orbit coupling can be adjusted by changing the symmetry of the quantum well containing the 2DEG by means of a gate electrode [10,11]. Many concepts of spin electronic devices rely on

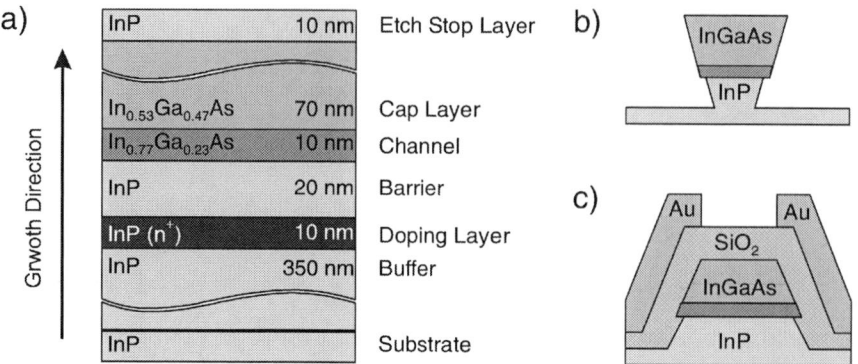

Figure 1. (a) Layer sequence of the InGaAs/InP heterostructure. (b) Schematics of the cross section of a wet chemically etched quantum wire oriented along $[0\bar{1}\bar{1}]$. (c) Cross section of a wet chemically etched wire aligned along $[0\bar{1}1]$. The split-gate electrodes are isolated from the semiconductor by a SiO_2 layer.

a reduction of the electron transport to only one dimension [5-7]. The effect of the Rashba effect on the transport in these one-dimensional conductors was theoretically investigated by Moroz and Barnes [12] and Mireles and Kirczenow [13]. Regarding experiments on wire structures it could be confirmed for relatively wide wires that the characteristic beating pattern appears in the magnetoresistance [14,15].

Here, we report on the Rashba effect in InGaAs/InP quantum wires. The wire structures were prepared by selective wet chemical etching. Down to a certain width the wires revealed a pronounce beating pattern in the magnetoresistance due to the Rashba effect. We will show that if the wire width is reduced further, the beating pattern vanishes completely. This fact is explained by the large energy separation of the magneto-subband due to the strong geometrical confinement. On a second set of samples which is supplied with split-gate electrodes, the strength of the Rashba effect was controlled by applying a gate voltage.

EXPERIMENTAL DETAILS

Our quantum wire structures are based on a InGaAs/InP heterostructure grown by metal-organic vapor phase epitaxy. The layer system is depicted in Fig. 1 (a). The 2DEG is located in the strained $In_{0.77}Ga_{0.23}As$ layer. The carriers are supplied by a Si-doped layer separated from the 2DEG by an InP barrier layer. The upper barrier is formed by an $In_{0.53}Ga_{0.47}As$ layer lattice-matched to InP. The top InP layer is used as an etch-stop layer for the subsequent wet chemical etching process.

The geometry of the quantum wires was defined by electron beam lithography. For the subsequent etching processes first a Ti mask was prepared by lift-off. Next, the pattern of the Ti mask was transferred into the upper InP layer by reactive ion etching (CH_4/H_2). Afterwards, the Ti mask is removed by HF/H_2O. The purpose of transferring the geometry into the top InP layer is to provide a tight mask for the subsequent wet chemical etching processes. The InGaAs layers are etched selectively by $H_3PO_4/H_2O_2/H_2O$. In final step the InP barrier and the InP dopant layer is wet chemically etched by using HCl/H_3PO_4. By this step the top InP etch-stop layer is also removed. As shown schematically in Fig. 1 b) and c), the cross section of the wires has a dovetail

G5.7.3

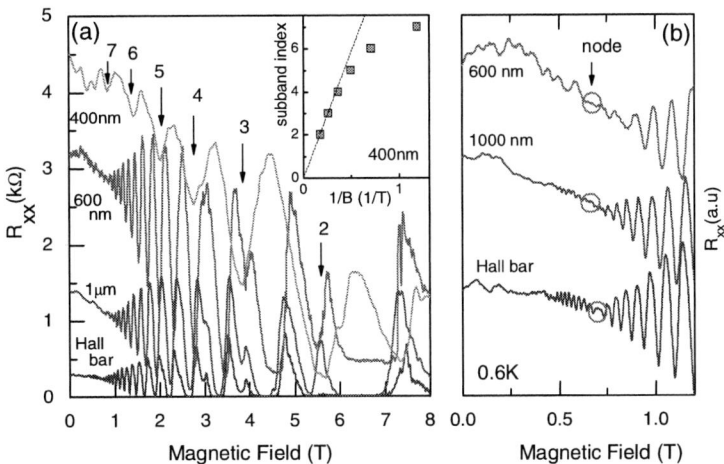

Figure 2. (a) Magnetoresistance of wires of various width aligned along the [01̄1̄] direction at T = 0.6 K. For comparison the Shubnikov-de Haas oscillations of a Hall bar are also shown. For the 400 nm wide wire the number of occupied subbands is indicated by arrows. The inset shows the number of occupied subband as a function of the inverse magnetic field. (b) Magneto-resistance of the wire structures at small magnetic fields.

shape if oriented along the [01̄1̄] direction and mesa-type for an orientation along [01̄1]. The nominal width of the wires, defined by the width of the Ti mask varied between 1μm and 400 nm. The effective width of the conductive layer is either smaller or larger by about 100 nm depending on the orientation of the wires (see figure 1). The sidewalls of the mesa-shaped wires were covered by split-gate electrodes, which were isolated from the semiconductor surface by a SiO_2 layer.

DISCUSSION

We will first discuss the magnetotransport of the dovetail-shaped wires having no gate electrode (figure 1 b). The magnetoresistance of wires of various width is shown in figure 2. As reference the Shubnikov-de Haas oscillations measured on a large-size Hall bar sample are also plotted. All wire structures reveal pronounced resistance oscillations as a function of the magnetic field B. Apart from the narrowest wire, the magnetoresistance oscillations are periodic in 1/B. By extracting the corresponding oscillation frequency f from a fast Fourier transform, the carrier concentration n_{2D} was determined by using the expression $n_{2D} = (2e/h) f$. It is found that the carrier concentration of the 1 μm wide wire agrees to the value of the Hall bar sample, whereas the carrier concentration of the 600 nm wire, being 5.2×10^{11} cm^{-2}, is smaller compared to the 1 μm wire. For the 400 nm wire the fast Fourier transform shows no well resolved peak, owing to the lower number of periods. However, from the period of the magnetoresistance in 1/B an even lower carrier concentration of 3.3×10^{11} cm^{-2} was estimated. A closer look on the resistance oscillations of the 400 nm wire reveals a clear deviation from the 1/B periodicity at

75

low magnetic fields. This can be seen in figure 2 (inset), where the subband occupation index is given as a function of the inverse magnetic field. Obviously, at low magnetic fields a deviation from the linear dependency, which is expected for a 2DEG, occurs. Thus, for the 400 nm wide wire the quantization due to the lateral confinement cannot be neglected.

A closer inspection of the magnetoresistance of the 1 μm and 600 nm wires reveals a characteristic beating pattern (figure 2 b) which can be attributed to the presence of Rashba effect. With decreasing width the node in the beating pattern shifts from 0.70 T for the Hall bar to 0.67 T for the 600 nm wire. As explained above, the Rashba effect leads to a separation into two spin-split subbands. Since the effective electron concentrations of the spin-split subbands are slightly different, a characteristic beating pattern is observed in the magnetoresistance. In case of the 1 μm wire and the Hall bar the resulting different oscillation frequencies lead to splitting of the peak in the fast Fourier spectrum. From the difference in the electron concentration Δn determined from the double-peak in the fast Fourier transform the parameter α_R can be extracted by using the following expression [11]:

$$\alpha_R = \frac{\Delta n}{\sqrt{2(n - \Delta n)}} \frac{\sqrt{\pi}\hbar^2}{m^*} . \tag{2}$$

Here, n denotes the average carrier concentration. For the Hall bar as well as for the 1 μm wide wire a Rashba spin-orbit coupling parameter of 7×10^{-12} eVm was determined. The node in the beating pattern of the 600 nm wire is found at almost the same magnetic field as the one of the 1 μm wide wire. Since the electron concentration of the latter structure is lower, it can be concluded that α_R is increased [11]. This increase can be attributed to the enhanced effective electric field in the quantum well, due to the stronger lateral confinement potential for decreasing wire widths. For the very narrow wire with a nominal width of 400 nm no beating effect is observed. This can be explained by the very large confinement energy. If we assume a harmonic confinement potential, a subband separation of 3.5 meV is estimated. This value is in the same order of magnitude as the typical Rashba energy splitting $2\alpha k_F$ at the Fermi level. The large level separation, due to the geometrical confinement, prevents the crossing of sublevels if the magnetic field is increased and thus suppresses the beating pattern in the magnetoresistance.

By using split-gate electrodes the carrier concentration in the wire can be controlled. This is shown in figure 3 a), where the magnetoresistance of a 400 nm wide mesa-shaped wire is plotted for various voltages applied to the gate. By decreasing the gate voltage towards negative values the carrier concentration is lowered, indicated by the increasing oscillation period. From the period in 1/B a carrier concentration of $n_{2D} = 8.5 \times 10^{11}$ cm^{-2} is calculated for a gate voltage of V_{gate}=+3 V which is reduced to 3.2×10^{11} cm^{-2} for V_{gate}=-2 V. A detail of the magnetoresistance at low magnetic fields is shown in figure 3 b). For this wire, the node in the beating pattern is found at higher magnetic fields. At V_{gate} = +3 V a node is observed at 1.6 T, whereas for V_{gate} = -1 V it is found at 1.2 T.

For the 400 nm mesa-shaped wire the fast Fourier transform of the magnetoresistance curves did not show any double-peak structure, so that α_R could not be extracted by the method described above. In order to estimate α_R, the subband spectrum of the wire was calculated numerically. Due to the small wire width, the confinement potential can be well approximated by a harmonic potential. As input parameter for the simulation the nominal width w and the carrier concentration n_{2D} determined from the magneto-oscillations at large magnetic fields were taken. From the value of n_{2D} the Fermi energy was determined. The sublevel spacing was calculated

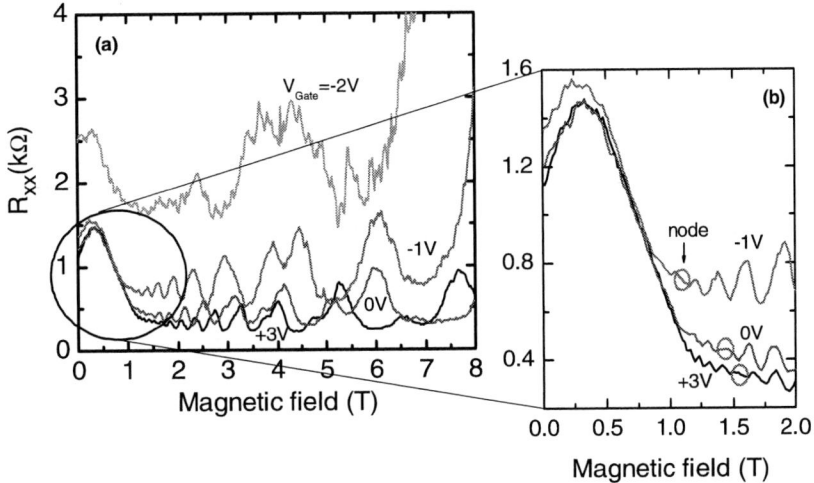

Figure 3. (a) Magnetoresistance of the 400 nm wide wire structure for various voltages applied to the split-gate. (b) Detail of the magnetoresistance oscillations at lower magnetic fields.

iteratively by filling the magnetosubbands up to the one-dimensional concentration $n_{1D}=n_{2D}\cdot w$ up to the given value of E_F. The position of the nodes in the beating pattern was determined from the intersection of the magneto-subbands with the Fermi level E_F. A node is expected at magnetic fields, where the spin-split subbands are equally spaced at E_F. The result of the simulation is shown in figure 4, where the expected node positions are indicated by the blue colored areas. It can be seen that a second node is expected theoretically, which is experimentally not resolved, due to the damping of the oscillation amplitude at low magnetic fields. By means of the graphs shown in figure 4, α_R can be determined for the experimentally observed node positions. For $V_{gate} = +3$ V a coupling parameter of 13.6×10^{-12} eVm was extracted, while at zero voltage α_R $=14.0 \times 10^{-12}$ eVm was found. For -2 V an even higher coefficient of 14.4×10^{-12} eVm was determined. The increase of α_R can be explained by the increased effective electric field in the wire [13]. Compared to the dovetail-shaped wires a considerably larger value of α_R is found. Due to the small wire width the potential shape within the wire should strongly depend on the interface at the edges of the wire. For the mesa-shaped wires the potential profile is probably modified by the presence of the SiO_2 layer, leading to a larger effective electric field and thus to an enhanced value of α_R.

CONCLUSIONS

In conclusion, for wider quantum wire structures the characteristic beating pattern can be observed similar to the beating found for a 2DEG. However, the node position can deviate largely compared to the value of the 2DEG, due to the contribution of the lateral confinement and due to interface effects at the sidewalls of the wires. For very narrow wires, with a large sublevel

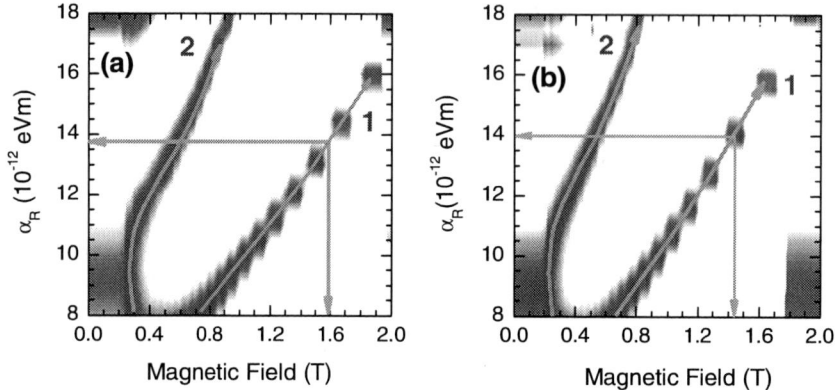

Figure 4. Calculation of the expected node position (blue areas) as a function of the B for different values of α_R. The arrows indicate the extracted value for the experimentally determined node positions. (a) Simulations for $n_{2D} = 8.5 \times 10^{11}$ cm^{-2} at $V_{gate}= +3$ V. (b) Simulations for $n_{2D} = 8.5 \times 10^{11}$ cm^{-2} at $V_{gate}= 0$ V.

separation, owing to the strong lateral confinement, no beating in the magnetoresistance is theoretically expected and experimentally observed.

REFERENCES

1. S. A. Wolf, *Science* **294**, 1488 (2001).
2. H. Ohno, *Science* **281**, 951 (1998).
3. E. I. Rashba, *Fiz. Tverd. Tela* (Leningrad) **2**, 1224 (1960) [Sov. Phys. Solid State **2**, 1109 (1960)]
4. S. Datta and B. Das, *Appl. Phys. Lett.*, **56** 665 (1990).
5. J. Nitta, F. E. Meyer, and H. Takayanagi, *Appl. Phys. Lett.* **75** 695 (1999).
6. A. A. Kiselev and K. W. Kim, *Appl. Phys. Lett.* **78,** 775 (2001).
7. M. Governale, D. Boese, U. Zülicke and C. Schroll, *Phys. Rev. B* **65**, 140403 (2002).
8. J. Luo, M. Munekata, F.F. Fang, and P. J. Stiles, *Phys. Rev. B* **38** 10142 (1988).
9. B. Das. D.C. Miller S. Datta, R. Reifenberger, W.P. Hong, P.K. Battacharya, J. Singh, and M. Jaffe, *Phys. Rev. B* **39** 1411 (1989).
10. J. Nitta, T. Akazaki, H. Takayanagi, and T. Enoki, *Phys. Rev. Lett.* **78**, 1335 (1997).
11. Th. Schäpers, G. Engels, J. Lange, Th. Klocke, M. Hollfelder, and H. Lüth, *J. Appl. Phys.* **83,** 4324 (1998).
12. A. V. Moroz and C. H. W. Barnes, *Phys. Rev. B* **60,** 14272 (1999).
13. F. Mireles and G. Kirczenow, Phys. Rev. *B* **64** 024426 (2001).
14. Y. Sato, S. Gozu, T. Kikutani, and S. Yamada, *Physica B* **272**, 114 (1999).
15. Th. Schäpers, J. Knobbe, A. van der Hart, and H. Hardtdegen, *Sci. Technol. Adv. Mat.* **4,** 19 (2003)

Mat. Res. Soc. Symp. Proc. Vol. 825E © 2004 Materials Research Society

Selective Voltage-Controlled Hole Spin in Non-Magnetic Resonant Tunneling Diodes

A. C. R. Bittencourt,[1] G. E. Marques,[2] Y. Galvão Gobato,[3] A.Vercik,[4] I. Camps,[5] and M. J. S. P Brasil[6]

[1]Departamento de Física, Universidade Federal do Amazonas, 67.077-000, Manaus, Amazonas. Brazil.

[2]Departamento de Física, Universidade Federal de São Carlos, 13565-905, São Carlos, São Paulo, Brazil

[3]Departamento de Física, Universidade Federal de São Carlos, 13565-905, São Carlos, São Paulo, Brazil, and Universidade Federal de Santa Catarina, Florianópolis, SC, Brasil.

[4]Departamento de Física, Universidade Federal de São Carlos, 13565-905, São Carlos, São Paulo, Brazil, and Faculdade de Zootecnia e Engenharia de Alimentos, Departemento de CIências Básicas, Pirassununga, 13635-900, SP, Brasil

[5]Departamento de Física, Universidade Federal de São Carlos, 13565-905, São Carlos, São Paulo, Brasil.

[6]Grupo de Propriedades Ópticas, Instituto de Física Gleb Wataghin, Universidade de Campinas, 13083-970, Campinas, SP, Brasil

(Dated: April 5, 2004)

We report theoretical and experimental observation of photoexcitated hole spin selection in GaAs/GaAlAs n-i-n in resonant tunneling diodes. When subjected to magnetic and electric parallel fields, the spin splitted hole levels leads to several peak structure in the transmissivity. These experimental results are interpreted as an evidence of tunneling transport through spin polarized hole levels of non-magnetic diodes.

PACS numbers: 71.70.Ej, 73.21.La, 78.30.Fs

Keywords: spin filter, resonant diodes, hole spin polarization

Resonant tunneling of holes in double barrier resonant tunneling structures (DBRTS) still attracts great attention due to both theoretical interest as well as technological applications in new areas such as spintronics. Whereas electron transport is mostly well understood, the complexity of the valence band structure gives rise to a variety of phenomena[1-4] that strongly influences all properties of a two-dimensional (2D) hole systems and deserves attention of researchers. The effects of a magnetic field on transport have been extensively studied and the magneto-tunneling have become a widely used spectroscopic technique.

In this work, we report the theoretical and experimental observation of valence-band Landau level structure in $n - i - n$ DBRTS, using photo-induced magneto-tunneling techniques. Here, the strong valence band admixture and the strong and nonlinear Zeeman splitting of hole carriers determine the resonant peak position as well as the overall general shape of the magneto-current in a given sample. The structures used in the work are symmetric $n - i - n$ GaAs/Al$_{0.35}$Ga$_{0.65}$As thick diodes with 10-nm i-Al$_{0.35}$Ga$_{0.65}$As barriers and a 5-nm i-GaAs well layers. The double barrier system is enclosed by a 60 nm i-GaAs and a 300nm Si-doped n$^+$-GaAs ($~10^{18}$cm^{-3}) layers grown on both sides of the structure. Annular contacts on the top of 500μm $\times$ 600μm mesas allow photocurrent measurements under applied voltage. The 488 nm line of a Coherent Ar$^+$ ion laser was used as photoexcitation source.

I. LANDAU-STARK HOLE STATES

The electronic structure of holes in a quantum well (QW) layer was proved by resonant magneto-tunneling spectroscopy and the formation of valence band levels have been experimentally studied by optical techniques in different heterostructures[4]. An important difference between conduction and valence tunneling is that the $I - V$ current display strong nonlinearities due to the crossing and anticrossing regions in the hole dispersions.

Let us consider homogeneous magnetic and electric fields $\mathbf{B} = (0, 0, B_0)$ and $\mathbf{F} = (0, 0, F_0)$, parallel to the z-axis of an isolated QW. This requires the replacement $\mathbf{k} = -i\boldsymbol{\nabla} + \frac{e}{\hbar c}\mathbf{A}$, where $\mathbf{A}$ is the vector potential in the Landau gauge. With magnetic field, the plane-wave (without B_0) solutions for momentum $\mathbf{k}_\rho = (\mathbf{k_x}, \mathbf{k_y})$ becomes cyclotron orbits with Landau envelope functions states, $|N_L> = |\varphi_{N_L,k_y}(\overrightarrow{\rho}) = e^{ik_y y}H_{N_L}(x - x_C)$, with quantum numbers $N_L = 0, 1, 2..$, and eigenvalues $E(N_L) = (N_L + 1/2)\hbar\omega_c$. Here $H_{N_L}(x - x_C)$ is Hermite function centered at $x_C = -(\hbar k_y/m^*\omega_c)$, $\omega_c = (eB_0/m^*c)$ is the cyclotron frequency, $\hbar k_y$ is the linear momentum component defining the orbit center x_C and $\ell_B = \sqrt{\hbar/(m\omega_c)}$ is the cyclotron radius. The electric field is treated as an uniform potential, $V(z) = -eF_0 z$, which is added to the offset profile of the heterostructure. Without electric field, states are composed of Bloch states at Γ−point and product of envelope function components, $A_\mu^m(z)$, $[\mu = hh^\sigma (lh^\sigma)$ for spin, $\sigma = \pm\frac{3}{2}(\pm\frac{1}{2})]$ and harmonic oscillator functions, $|N_L>$. One realistic model to treat the valence band states is the well known Luttinger-Kohn Hamiltonian model[5], that contains a considerable degree of mixture and provides a very precise description for those semiconductors where the spin-orbit (Δ_0), and the band-gap (E_g) energies are large (say E_g, $\Delta_0 \geq 300$ meV).

The energy and state of light-holes $(lh_{N_L}^\sigma)$ or heavy-holes $(hh_{N_L}^\sigma)$ for given Landau number (N_L) and spin-polarization (σ) can be determined from solutions of $H_{Lutt}\Psi_{\mu^\sigma}^m(z,\rho,N_L) = E_{\mu^\sigma}^m(B_0,F_0)\Psi_{\mu^\sigma}^m(z,\rho,N_L)$ in the form

$$
\begin{bmatrix}
P_{hh}^+ & L_1 & 0 & M \\
L_1^\dagger & P_{lh}^+ & M & 0 \\
0 & M^\dagger & P_{hh}^- & L_1^\dagger \\
M^\dagger & 0 & L_1 & P_{lh}^-
\end{bmatrix}
\begin{bmatrix}
A_{hh+}^m(z)|N_L-2> \\
A_{lh+}^m(z)|N_L-1> \\
A_{hh-}^m(z)|N_L+1> \\
A_{lh-}^m(z) \quad |N_L>
\end{bmatrix}
= E_{\mu^\sigma}^m(B_0,F_0)
\begin{bmatrix}
A_{hh+}^m(z)|N_L-2> \\
A_{lh+}^m(z)|N_L-1> \\
A_{hh-}^m(z)|N_L+1> \\
A_{lh-}^m(z) \quad |N_L>
\end{bmatrix}
\tag{1}
$$

where $\mu^\sigma = hh_{N_L}^\sigma$, $lh_{N_L}^\sigma$ labels the polarized carrier type and Landau level. The elements defining this matrix are: $P^\pm{}_{hh} = V(z) + [(\gamma_1 - 2\gamma_2)/2]\frac{\partial^2}{\partial z^2} - S(B_0)[(\gamma_1 + \gamma_2)/2\,(2\hat{n}+1) \mp \frac{3}{2}(\kappa + \frac{9}{4}q)]$; $P^\pm{}_{lh} = V(z) + [(\gamma_1 + 2\gamma_2)/2]\frac{\partial^2}{\partial z^2} - s(B_0)[(\gamma_1 - \gamma_2)/2\,(2\hat{n}+1) \mp \frac{1}{2}(\kappa + \frac{1}{4}q)]$; $L_1 = -i(\sqrt{6}/\ell_B)\gamma_3\frac{\partial}{\partial z}\hat{a}$; and $M = \sqrt{3}\,s(B_0)[(\gamma_2 + \gamma_3)/2\,(\hat{a})^2 + (\gamma_2 - \gamma_3)/2\,(\hat{a}^+)^2]$. In these expressions, γ_1, γ_2, γ_3 are the Luttinger-Kohn-Kane effective mass parameters, κ and q are magnetic Luttinger-Kohn[5] parameters defining the hole g-factor (all measured in units of $\frac{\hbar^2}{m_0}$), $s(B_0) = \hbar\omega_c = 0.115764\,B_0$ is the cyclotron energy (in meV) for B_0 given in Tesla. Also, $V(z) = V_0(z) - eF_0 z$ is the full potential that includes the band-offset profile, $V_0(z)$, for the materials in QW interface. Yet, $\hat{n} = \hat{a}^+\hat{a}$ is the oscillator number operator, $\hat{n}\,|N_L> = N_L\,|N_L>$, for each Landau state $|N_L>$. The diagonal terms identify the ordering of carrier whereas the Landau numbers in each spinor component were dictate after setting $\gamma_2 = \gamma_3$ only on the warping term, proportional to $(\hat{a}^+)^2$ in the element M. Anywhere else we have kept the axial form of this Hamiltonian with $\gamma_2 \neq \gamma_3$. We solve this multiband effective mass equation by using finite-difference method.

In the absence of both fields, each QW function is either symmetric (even), $A_{hh\pm(lh\pm)}^m(z)$ $m = 1,3,5..$, or anti-symmetric (odd), $A_{hh\pm(lh\pm)}^m(z)$ $m = 2,4,..$ at $B_0 = 0$ (or $\mathbf{k}_\rho = 0$). Only in this single case we can call these solutions as pure heavy-hole or light-hole states with the degenerated Kramer doublets, $lh^\pm$ or $hh^\pm$. At $B_0 \neq 0$, there is a Zeeman separation between spin-up $(lh^+$ or $hh^+)$ and spin-down $(lh^-$ or $hh^-)$ carriers for different Landau numbers. This strong mixing, inherent to $\mathbf{k.p}$ model, gives origin to a highly anomalous g-factors as well as different cyclotron effective masses for each hole spin polarized state. The further application of an electric field, F_0, induces mixture of parities for different components of spinors and, thus, breaks the inversion symmetry of the envelope functions $A_{\mu^\sigma}^m(z)'s$. This broken time-reversal symmetry adds another contribution to the spin splitted Kramer doublets. The combined Landau-Stark features are very important in order to understand any transport or optical property of a system. For each state we can observe anomalous g-factors $(g_{hh_{N_L}(lh_{NL})}(B_0,F_0))$ as well as different cyclotron effective masses $(m_{hh_{N_L}(lh_{NL})}^\pm(B_0,F_0))$, depending on both field strengths and quantum numbers N_L.

After a careful analysis of the possible solutions, we have found that the mixing in the valence band, as determined by the combination of the operators $\frac{\partial}{\partial z}$ and $\hat{a}$ in the off-diagonal elements of H_{Lutt}, imposes condition on each Landau-Stark state construction. The sequence of states $(|NL>)$, occurring in each component of a spinor shown above as well as the parity of z-dependent envelope functions $(A_{hh\pm(lh\pm)}^m(z))$ is a direct consequence of this admixture. A deep look into the structure of their Hilbert space shows that there are four sets of Landau-Stark states with different combination of quantum numbers involved. Let us label each individual carrier type by the Landau number, n_L, for every possible value of the index, N_L, in component of the spinors indicated in Eq. [1].

i) The lowest possible value for the index is, $N_L = -1$. In the case only heavy-hole spin-down (hh_0^-) solutions with Landau number $n_L = 0$, can occur for each level $m = 1,2,3...$ inside the QW. They will display parabolic dispersions, $E_{hh_0^-}^m(B_0,F_0)$ (linear in B_0) and we have adopted the notation $\Psi_{hh-}^m(z,\rho,0)$ to refer to 1-component state. It is clear that these parabolic solutions, eigenvalues of the diagonal term P_{hh}^- in Eq. [1], does not couple to any other set of states.

ii) The next value $N_L = 0$ will determine a 2x2 Hamiltonian matrix coupling only hh_1^- and lh_0^- spin-down states. Their energy dispersions with respective Landau numbers occur in the construction of their states, $\Psi_{hh-}^m(z,\rho,1)$ and $\Psi_{lh-}^m(z,\rho,0)$. These two states, for each value of m, interact with each other and their dispersions, $E_{hh_1^-}^m(B_0,F_0)$, $E_{lh_0^-}^m(B_0,F_0)$ can show anticrossing regions for special values of B_0. A small energy separation, in order to observe minigap regions for these two states, will depend on the relative effective masses, QW width and, mainly, on the applied field F_0.

iii) For the next value, $N_L = 1$, we must find solutions for hole states described by a 3x3 matrix coupling hh_2^-, lh_1^- and lh_0^+ states. Their spinor form are written as: $\Psi_{hh-}^m(z,\rho,2)$, $\Psi_{lh-}^m(z,\rho,1)$ and $\Psi_{lh+}^m(z,\rho,0)$. Notice the first occurrence of lh_0^+ in this multiplet that interacts with its partner states. The same conditions for minigap regions can be extended here for $lh's$ and $hh's$ states, however lh_0^+, lh_1^- can display much stronger interaction (for same value of m), at special value of B_0.

iv) In the next sequences, with indexes $N_L \gtrsim 2$, all four spin-down and spin-up components of the carriers are coupled and the individual Landau numbers appear in the form $\Psi_{hh+}^m(z,\rho,N_L-2)$, $\Psi_{hh-}^m(z,\rho,N_L+1)$, $\Psi_{lh+}^m(z,\rho,N_L-1)$

and specific profile $V_0(z)$. The layer $\ell = 0$ ($\ell = 4$) is the emitter (collector) contact. We use the $S-$matrix technique to calculate the *transmissivity* and *reflectivity* on the system under applied voltage (electric field). For a given incident energy (E) of a carrier, we construct *incoming* and *outgoing* spinor states that will tunnel through the 2D hole states in the well layer $\ell = 2$. The main steps can be summarized as: i) The envelope function for an carrier in position $\mathbf{r} = [\boldsymbol{\rho} = (\mathbf{x}, \mathbf{y}), z]$, with energy E, propagating linear momentum $\hbar k_z$ and spin-polarization σ, is constructed as linear combination of *travelling waves* of the perfect crystal (bulk). From the emitter ($l = 0$) to the collector ($l = 4$) layers, each spinor component, in Eq. (??), has the form[6]

$$F(z) = \sum_{\mu = lh, hh} \sum_{\sigma = (+, -)} [a_\mu^\sigma(l) \mathbf{F}_{\mu, \sigma}(+k_{z\sigma}) e^{+ik_{z\mu}^\sigma(z - z_l)} + b_\mu^\sigma(l) \mathbf{F}_{\mu, \sigma}(-k_{z\sigma}) e^{-ik_{z\mu}^\sigma(z - z_l)}], \qquad (2)$$

where $a_\mu^\sigma(l)$ ($b_\mu^\sigma(l)$) are the amplitudes of *incoming* (*outgoing*) polarized hole waves at the interface, z_l, and $\pm k_{z\mu}^\sigma$ are *real* roots of $E_{\mu^\sigma}(B_0, k_z) = E$, for each wave travelling to positive ($+k_{z\mu}^\sigma$) and negative ($-k_{z\mu}^\sigma$) direction along the $z-$axis, respectively. Here $E_{\mu^\sigma}(B_0, k_z)$ are heavy- and light-hole "bulk eigenvalues" for the Luttinger-Kohn Hamiltonian; ii) The usual boundary conditions requires that both matrices $\boldsymbol{\Psi}_{\mu^\sigma}(\rho, z = z_l, N_L)$ and carrier flux $\mathbf{J}_z \boldsymbol{\Psi}_{\mu^\sigma}(\rho, z = z_l, N_L)$ be continuous across each interface z_l. The dimension of these matrices changes according to which N_L multiplet is being calculated. Yet, resonant magneto-tunneling without any scattering implies that these interface matching conditions will be performed under n_L-Landau number conservation. Furthermore, the current density operator, $\mathbf{J}_z$, for Hamiltonian in Eq. [1], has the general 4×4 matrix form

$$\mathbf{J}_z(B_0, k_z) = \begin{bmatrix} (\gamma_1 - 2\gamma_2) \frac{\partial}{\partial z} & \sqrt{6}\, \gamma_3\, \hat{a}/\ell_B & 0 & 0 \\ \sqrt{6}\, \gamma_3\, \hat{a}^\dagger/\ell_B & (\gamma_1 + 2\gamma_2) \frac{\partial}{\partial z} & 0 & 0 \\ 0 & 0 & (\gamma_1 - 2\gamma_2) \frac{\partial}{\partial z} & \sqrt{6}\, \gamma_3\, \hat{a}^\dagger/\ell_B \\ 0 & 0 & \sqrt{6}\gamma_3\, \hat{a}/\ell_B & (\gamma_1 + 2\gamma_2) \frac{\partial}{\partial z} \end{bmatrix}, \qquad (3)$$

where the diagonal (off-diagonal) terms represent the spin and carrier type conserving (carrier type flipping) mechanisms inherent to Luttinger-Kohn Hamiltonian[6]. It is clear that eight types of spin-polarized current can be established in the sample, according to the incident (injected) polarization (*up* and *down*) of a carrier type (hh or lh) and to the transmitted (detected) polarization and type at the collector. The ratio between *wave amplitudes* at the emitter ($l = 0$) and collector ($l = 4$) define four elements in the 8×8 S-matrix, $t_\sigma^{\sigma'}(\mu) = a_\mu^{\sigma'}(4)/a_\mu^\sigma(0)$, $\mu = lh, hh$; $\sigma, \sigma' = +, -$, or the so called *transmission coefficients*. The other four elements are the *reflection coefficients*, $r_\sigma^{\sigma'}(\mu) = a_\mu^{\sigma'}(4)/a_\mu^\sigma(0)$, $\mu = lh, hh$; $\sigma, \sigma' = +, -$. These terms are used to calculate the partial *trasmissivity* ($T_{\sigma \to \sigma'}$) for each incident(σ) detected (σ') polarization as:

$$T_{\sigma \to \sigma'}(E, V, B_0) = \Re e[\frac{\mathbf{J}_{\sigma'}^{out}}{\mathbf{J}_\sigma^{in}}], \qquad (4)$$

or, the ratio between the current detected(out) in σ'-polarization ($\mathbf{J}_{\sigma'}^{out} = \sum_{\mu'} \{[t_\sigma^{\sigma'}(\mu)]^* t_\sigma^{\sigma'}(\mu)] < \mathbf{F}_{\mu, \sigma'}^+|\mathbf{J}_{z\sigma'}^+|\mathbf{F}_{\mu, \sigma'}^+ >\}$) to the incident(in) with $\sigma-$polarization ($\mathbf{J}_\sigma^{in} = \sum_\mu < \mathbf{F}_{\mu, \sigma}^+|\mathbf{J}_{z\mu, \sigma}^+|\mathbf{F}_{\mu, \sigma}^+ >$). Here we use a shorter notation, $\mathbf{F}_{\mu, \sigma}(\pm k_{z\sigma}) \equiv \mathbf{F}_{\mu, \sigma}^\pm$ and $\mathbf{J}_{z_\mu}(B_0, \pm k_{z\mu}^\sigma) \equiv \mathbf{J}_{z\mu, \sigma}^\pm$. The partial *reflectivity* (R) can be derived from flux conservation, $\sum_{\sigma'}[T_{\sigma \to \sigma'}(E, V, B_0) + R_{\sigma \to \sigma'}(E, V, B_0)] = 1$, for each fixed incident $\sigma-$polarization. Further details on this approach can be found in Ref.[6]. In principle, there will be transmissivity peaks every time a spin-polarized QW hole level becomes resonant with incident energy below the Fermi level (E_F) at the emitter(injector) layer. These transmissivities measure the tunneling probabilities. Besides, we must assert an occupied (empty) energy state at the emitter (collector)layer. The full current is obtained as[6]

$$J_\sigma(B_0) = \sum_{n_L}^{N\max(E)} \int_{E_F - eV}^{E_F} T_{\sigma \to \sigma'}(E, V, B_0) \Delta F(E, V) D(B_0, E) dE, \qquad (5)$$

where $\Delta F(E, V) = [F(E) - F(E - eV)]$ is the joined Fermi probability assuring the existence of an occupied (empty) state in the emitter (collector) side, V ($-eV$) is the voltage (electrostatic potential energy) drop across the sample and $N_{\max}(E)$ is the largest occupied Landau level that can be determined from the condition, $E \geq \{E_{\mu_{N\max(E)}^{\sigma'}}^m (B_0, F_0)\}$ for each carrier $\mu = lh, hh$ and spin σ'. The simplest example is for parabolic Hamiltonian where, $E_{n_L}(B_0) = \hbar\omega_c (n_L + \frac{1}{2})$. For this case, the condition determining $N\max(E)$ becomes $E \geq \{\hbar\omega_c[N_{\max}(E) + 1/2]\}$[7]. Finally, $D(B_0, E)$ is the density of states for any chosen dimensionality to treat the dispersions on collector terminal. In order to simplify the integration on the current, we will consider as 3D terminal, then $D(B_0, E) = (\frac{e^2 B_0}{4\pi^2 \hbar^2 c})$.

G5.10.5

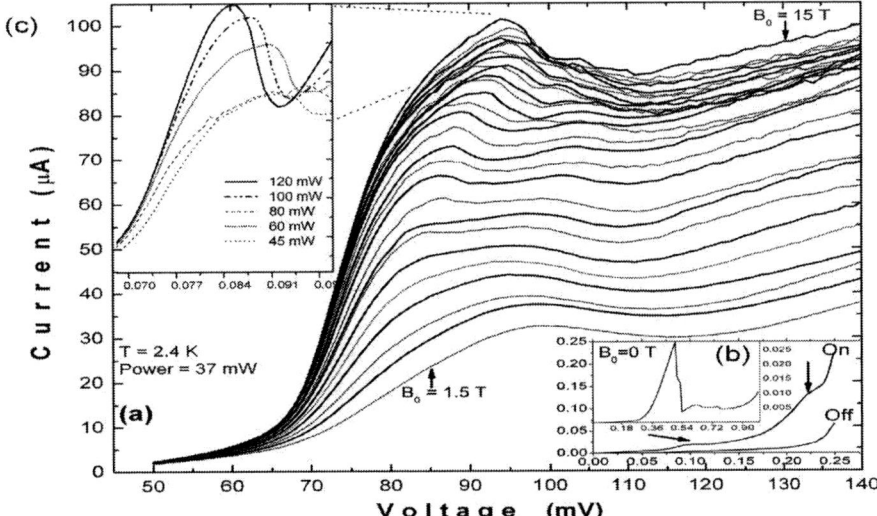

FIG. 1: Part a): As field increases from 1.5 to 15 T, the broad peak shows hidden structures resonant to hole different Landau-Stark ground-states ($m = 1$). Inset b): Current in dark (Off) and illuminated (On), showing hole structures for $m = 1$ and $m = 2$, at $B_0 = 0$. Inset c): Effect of excitation power on the lowest structure.

II. RESULTS

The inset b of Fig. 3 shows the dark current (Off) measured in the $n - i - n$ sample which is only determined by electrons. At the onset voltage (< 0.25 V) of the electron current, no effect of the magnetic field is observed.

When the sample is illuminated (On), without magnetic field, an appreciable increase of current is observed in this low voltage range (inset b). The two observed small and broad peaks are associated to resonant current of photocreated holes on the top mesa contact and driven by applied voltage. The identical amount of electrons photogenerated does not affect the electron current (inset b) in this n^+ symmetric doped sample. Any current-voltage ($I - V$) curve here will always include both electron and hole contributions thus, the optical excitation power was kept low to avoid an earlier (at lower voltage) increase of the electron current which would screen-out or mask the hole current contribution in our sample. All measurements shown in Fig.3 were taken at 2.4 K with field provided by an Oxford superconducting magnet, changing from 1.5 to 15T, in steps of 0.5 T.

The observed sequence of peaks corresponds to resonant tunneling of holes with Landau-Stark levels produced by the combination of both electric and magnetic parallel fields.

The calculated current of holes in the sample, assumes an uniform potential drop along the sample. There is a difference between calculated and experimental reality, since the sample is n-doped and the voltage drop at each layer could only be known from a selfconsistent calculation for specific charge distribution. Therefore, there is an unknown difference between measured voltage (voltmeter) and calculation model. Our analyze seems to indicate only a simple overall shift near 40-50 mV between theoretical and experimental values.

Fig 4 shows the calculated current for the multiplet $N_L = 4$, when the heavy-holes are injected on the emitter side, with a density associated to a 5 meV quasi-Fermi Level (E_F) for holes and detected in both lh $spin$-up (dashed line) and $spin$-$down$(solid line) polarizations (panel a). In panel b, we are showing this current when detected on both hh $spin$-up and $spin$-$down$ polarizations. The apparent differences are produced by small changes on the transmissivities $T_{\sigma \to \sigma'}$ when using direct and opposite polarization and by the integration region in Eq.[5]. This is an interesting property if *magnetic detector mask* could be mounted on the collector side, in order to establish a filter selecting the magnetic polarization of the arriving carrier. This technology is possible[8] and is been explored by some groups in order to produce spin filters. If the detector cannot distinguish the spin polarization, we must add all contributions to the current from panels a and b, as shown in panel c (dashed and solid lines), together with the contribution from hh_0^- state (dotted lines) resonant with the first ($m = 1$) and second ($m = 2$) eigenvalues in the QW and presented as case $N_L = -1$. Therefore, almost all peaks occurring below 100 mV, are related to resonances with $m = 1$ sets of

83

G5.10.6

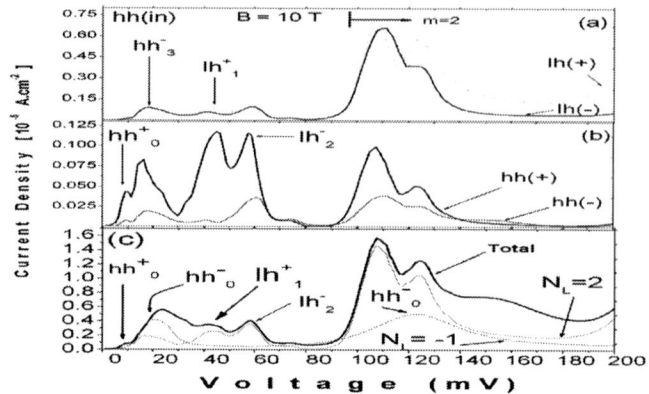

FIG. 2: Parts a and b show current for incident hh detected on both $lh^{\pm}$ and $hh^{\pm}$ polarizations. Part c shows the total currrent. The $hh_0^{\pm}$ doublet shows large spin polarization.

Landau-Stark $hh_{n_L}^{\sigma}$ and $lh_{n_L}^{\sigma}$ levels. Above 100 mV we observe another peak structure formation coming from sets with $m = 2$. They appear stronger since, besides the different scales, are closer to the offset potential and, thus, their tunneling becomes enhanced ($T_{\sigma \to \sigma'} \to 1$) for increasing voltage (or electric field).

The theoretical model shows a large degree of polarization, $P = [I(hh_0^+) - I(hh_0^-)]/[I(hh_0^+) + I(hh_0^-)]$ for the heavy-hole ground-state. These peaks occur at different voltages and, therefore, could be used as voltage controlled[9] spin-hole filter in this type of non-magnetic diode. The experimental curves show identical peak structures at displaced position(as mentioned above) but with a much smaller peak-to-valley ratio for this specific sample. Furthermore, the addition of other contributions to the theoretical curves will cause a broader peak below 80 mV, as can be drawn from the number Landau-Stark levels in Figs. 3 and 4. The other effect that should be considered is the power excitation that may increase the density of holes and the Fermi Level at the emitter, as shown on the inset c of Fig. 3, we show how the excitation increase affects the shape of this broad region where these peaks occur.

In summary, we reported the direct observation of Landau-Stark levels formation in resonant tunneling diodes. We discussed the nonlinear behavior of the g-factors, effective masses and how crossings and anti-crossings between different states are enhanced by electric and magnetic fields. The resonant hole energy was used to calculated polarized and unpolarized currents. We are able to interpret several aspects of the highly complex experimental curves based on our theoretical model. Finally, we showed that nonmagnetic diodes can work as voltage controlled spin-hole filters in specifically designed samples and these are under study in our group.

Work is suported by agencies FAPESP and CNPq.

---

[1] R.K. Hayden et al., Phys. Rev. Lett. 66, 1749 (1991).

[2] A. Zaslavsky, D.A. Grützmacher, S.Y. Lin, T.P. Smith III, R.A. Kiehl and T.O. Sedwick, Phys. Rev . B 47, 16036 (1993).

[3] E.E. Mendez, L. Esaki and W.L. Wang, Phys. Rev. B 33, 2893 (1986).

[4] R.K. Hayden, D.K. Maude, L. Eaves, E.C. Valadares, M. Henini, F.W. Sheard, O.K. Hughes, J.C. Portal and L. Cury, Phys. Rev. Lett. 66, 1749 (1991).

[5] J.M. Luttinger and W. Kohn, Phys. Rev. B 8, 2697 (1973); J.M. Luttinger, Phys. Rev. 102, 1030 (1955).

[6] A.C. Bittencourt, A.M. Cohen and G.E. Marques, Phys. Rev. B 57, 4525 (1998).

[7] C.B. Duke, "Tunneling In Solids", Solid State Physics, eds. F. Seitz, D. Turnbull, and H. Ehrenreich, Academic Press, 1969

[8] E.I. Rashba, Phys. Rev. B **62**, R16267 (2000).

[9] A. Slobodskyy, C. Gould, T. Slobodskyy, C.R. Beker, G. Schmidt, and L.W. Molenkamp, Phys. Rev. Lett. 90(24), 24601 (2003).

Mat. Res. Soc. Symp. Proc. Vol. 825E © 2004 Materials Research Society G6.6.1

SPIN LIFETIME TUNING IN ZINCBLENDE HETEROSTRUCTURES AND APPLICATIONS TO SPIN DEVICES

X. CARTOIXÀ[*], D. Z.-Y. TING[†], Y.-C. CHANG
Department of Physics, University of Illinois at Urbana-Champaign, Urbana, IL 61801, USA
[†] Jet Propulsion Laboratory, California Institute of Technology, Pasadena, CA 91109, USA

ABSTRACT

We present analytical expressions for the D'yakonov-Perel' spin relaxation rates under the combined action of bulk and structural inversion asymmetry for [111] zincblende heterostructures when terms up to linear and third order in k are included in the Hamiltonian. We see for [111] heterostructures that, under the right conditions, the lowest-order-in-k component of the spin relaxation tensor can be made to vanish for *all* spin components at the same time. We study how the inclusion of terms of higher order in k affects these results. We finally discuss a proposal for a resonant spin lifetime transistor (RSLT) using the spin lifetime tuning concepts presented above, where the characteristics of the [111] device give the designer an added degree of freedom on the direction of the injected spins.

1. INTRODUCTION

If the current pace of electronic device miniaturization is to continue, it is reasonable to think that the good use of the quantum properties of the electron will play a role in making this possible. Traditionally it has been the wave character of the electron that has been put to this use, resulting in devices such as the resonant tunnel diode [1] and the single electron transistor [2].

Another quantum property of the electron that only recently has received attention for its potential for information storage and processing is its spin. One of the key parameters that must be controlled for the successful achievement of *spin* elec*tronic* (spintronic) devices, such as the Datta-Das transistor [3], is the spin lifetime of the carriers. For the transport of spin-encoded quantum (single state) or classical (average over an ensemble) information, we naturally demand spin lifetimes as long as possible. If, as it is normally the case, we are to employ heterostructures in the design of our spintronic devices, we need tools that provide us with predictions about the spin lifetimes and direct us to ways of obtaining the goal of long spin lifetimes. The lack of an inversion symmetry center lifts the double degeneracy at a general k point in the Brillouin zone, thus greatly reducing the spin lifetime of electrons.

In this paper we investigate how the interplay of structural inversion asymmetry [4] (SIA) and bulk inversion asymmetry [5] (BIA) affects the D'yakonov-Perel'-Kachorovskiǐ [6, 7] spin lifetimes for electrons in [111] quantum wells (QWs). The effects of SIA on the spin dynamics should always be kept in mind, as it can be unintentionally present in any heterostructure due to uneven doping profiles [8], surface effects, different interdiffusion at the boundaries, etc. We start, in Sec. 2, by computing the effective spin Hamiltonian in a two-band model. In Sec. 3 we then proceed to compute the ensemble lifetime of the three spin components as a function of the relative magnitude of BIA and SIA contributions following the procedure from Refs. [9] and [10]. Finally, in Sec. 4 we review a newly proposed family of devices [11, 12] based on the special properties of the spin lifetime tensor when the BIA and SIA effects have equal strength in a [001] QW, as

85

pointed out by Averkiev and Golub [13], and Kiselev and Kim [14]. The same kind of devices has also been proposed in [110] structures by Hall et al. [15]. We show how [111] versions of the device are expected to have properties that make them easier to implement than their [001] and [110] counterparts.

2. TWO-BAND HAMILTONIANS

Here we present the effective two-band [spin-resolved conduction band (CB)] Hamiltonian corresponding to zincblende [111] QWs. We start from the $\mathcal{O}(k^3)$ spin part of the Hamiltonian for bulk zincblendes [6]

$$H_{\text{BIA}} = \gamma \left[\sigma_x k_x \left(k_y^2 - k_z^2 \right) + c.p. \right], \tag{1}$$

where σ_i are the Pauli matrices, k_i are the electron wavevector components and $c.p.$ stands for the cyclic permutation needed to obtain the remaining terms of the Hamiltonian.

We first do a change of basis to express H_{BIA} in natural coordinates for the [111]-grown structures. Then, following the procedure in Refs. [7] and [16], we quantize $\mathbf{k}$ along the growth direction and, keeping only terms linear in $\mathbf{k}_\parallel$—second order terms in $\mathbf{k}_\parallel$ vanish because of time reversal requirements for the expectation value of k_z [16]—, we arrive at the following expressions for the BIA Hamiltonian of [111] QWs

$$H_{\text{BIA}\,[111]} = \frac{2\gamma \langle \hat{k}_z^2 \rangle}{\sqrt{3}} \left(k_y \sigma_x - k_x \sigma_y \right), \tag{2}$$

where the labels x, y, z depend on the orientation of the structure.

Upon inspection of the Hamiltonian in Eq. (2) we see that BIA causes a k-dependent effective magnetic field pointing in-plane for [111] structures. Note that $H_{\text{BIA}\,[111]}$ is formally identical to the Rashba Hamiltonian [4]

$$H_{\text{R}} = \alpha_{\text{R}} \left(k_y \sigma_x - k_x \sigma_y \right), \tag{3}$$

where α_{R} is the Rashba coefficient, whose value depends on the particulars of the structural asymmetry present in the sample. We shall now see that this has important consequences in the values of the spin lifetimes.

3. SPIN LIFETIMES

Here we follow the methods of Averkiev and Golub [10, 13] to compute the spin lifetime of electrons in the CB of [111] QWs. The results presented are explained in more detail in Ref. [17].

The combination of Eqs. (2) and (3) yields the first order Hamiltonian

$$\begin{aligned} H_{\text{IA,1}} &= \left(\alpha_{\text{BIA}} + \alpha_{\text{R}} \right) \left(k_y \sigma_x - k_x \sigma_y \right) \\ &= \alpha_{\text{IA}} \left(k_y \sigma_x - k_x \sigma_y \right), \end{aligned} \tag{4}$$

where $\alpha_{\text{BIA}} \equiv 2\gamma \langle \hat{k}_z^2 \rangle / \sqrt{3}$ and we have introduced $\alpha_{\text{IA}} = \alpha_{\text{BIA}} + \alpha_{\text{R}}$ describing the combined effects of BIA and SIA in the heterostructure.

Since Eq. (4) is formally identical to the Rashba Hamiltonian, the existing results for SIA only [10, 13] will hold taking $\alpha_{\text{R}} \to \alpha_{\text{IA}}$:

$$\tilde{\tau}_x = \tilde{\tau}_y = 2\tilde{\tau}_z = \frac{\hbar^2}{2\alpha_{\text{IA}}^2} \frac{1}{k^2 \tilde{\tau}_1}, \tag{5}$$

where the tilde indicates a magnitude that is evaluated at a given energy and $\tilde{\tau}_1$ is the effective time for field reversal due to the harmonic $l = 1$ of the scattering cross section, and in general [10, 18] $\tilde{\tau}_l^{-1}(E) = \oint \sigma(\phi, E)(1 - \cos l\phi) d\phi$. The spin directions will be perpendicular to the wavevector and in-plane [19, 20]. We see that, as usual in the D'yakonov-Perel (DP) mechanism, the spin lifetime is inversely proportional to the momentum lifetime.

A most interesting configuration for [111]-grown samples occurs when $\alpha_{\text{BIA}} = -\alpha_{\text{R}}$. Then, $\alpha_{\text{IA}} = 0$ and the conduction bands become spin degenerate to first order in k. The key consequence of this configuration is that the spin lifetimes would be extended for *any* spin direction, as opposed to spins along [110] for (100) structures and $\alpha_{\text{BIA}} = \alpha_{\text{R}}$ [13] or spins perpendicular to the plane well for (110) structures and $\alpha_{\text{SIA}} = 0$ [7]. Control of α_{R} can be achieved by the application of a gate bias [19, 21] or by sample design with compositional asymmetry, providing a nonzero α_{R} at zero bias. Thus, properly biased (111) QWs could act as spin reservoirs, or form the basis of a resonant spin lifetime transistor as described in the next section.

If we include terms of order in k^3 in the Hamiltonian, we obtain the following results for the spin lifetimes [17]

$$\tilde{\tau}_x = \tilde{\tau}_y = \frac{6\hbar^2}{k^2 \tilde{\tau}_1} \frac{1}{12\alpha_{\text{IA}}^2 - 4\sqrt{3}\gamma\alpha_{\text{IA}}k^2 + (1 + 2\tilde{\tau}_3/\tilde{\tau}_1)\gamma^2 k^4}$$

$$\tilde{\tau}_z = \frac{3\hbar^2}{k^2 \tilde{\tau}_1} \frac{1}{\left(\gamma k^2 - 2\sqrt{3}\alpha_{\text{IA}}\right)^2}. \tag{6}$$

Since the scattering rate is proportional to $(H_{\text{IA}})^2$, the k^6 terms in Eq. (6) are not correct in general because terms arising from the combination of $H_{\text{IA},1}$ with fifth order contributions to H_{IA} are missing. However, we have kept the k^6 terms here because they are correct in the special case where $\alpha_{\text{IA}} = 0$, giving the lowest order contribution to the spin scattering rate.

Equation (6) is plot in Fig. 1 for typical values $k_F = 0.01 \text{ Å}^{-1}$, $\tau_p = 1$ ps, $\gamma = 186 \text{ eV·Å}^3$ [22], $\alpha_{\text{BIA}} = 11 \times 10^{-10}$ eV·cm [23] as a function of the ratio $\alpha_{\text{R}}/\alpha_{\text{BIA}}$. The τ_x, τ_y and τ_z components show the predicted resonant spin lifetime when $\alpha_{\text{R}} = -\alpha_{\text{BIA}}$. Although Eq. (6) limits by itself the values of the lifetimes at the resonance, these values are very large and, thus, other mechanisms [24–26] will effectively limit the value of the resonant spin lifetime. Therefore, [111]-grown heterostructures provide DP suppression on par with [110]-grown structures, with the added advantage the suppression is for *all* spin components, as opposed to one component only. We can estimate the value of the non-DP-limited spin lifetime by noting that in [110] structures spins along the growth axis have a lifetime of the order of ns [27, 28], larger than the tens of ps observed in [001] structures [27].

4. DEVICES

In what follows, we will describe how, by driving the spin lifetime in and out of resonance through the action of an external bias, we can construct a number of spintronic devices. Figure 2 shows the operating principle of the [111] resonant spin lifetime transistor (RSLT). The device layout is very similar to the Datta-Das [3] device. As opposed to the [001] ([110]) version of the device, where the ferromagnetic contacts must be designed so that their magnetization points in the [$\bar{1}$10] ([110]) direction, the fact that all spin components are resonant at the same time gives the designer freedom to choose the orientation of the magnetization. Thus, we can choose to

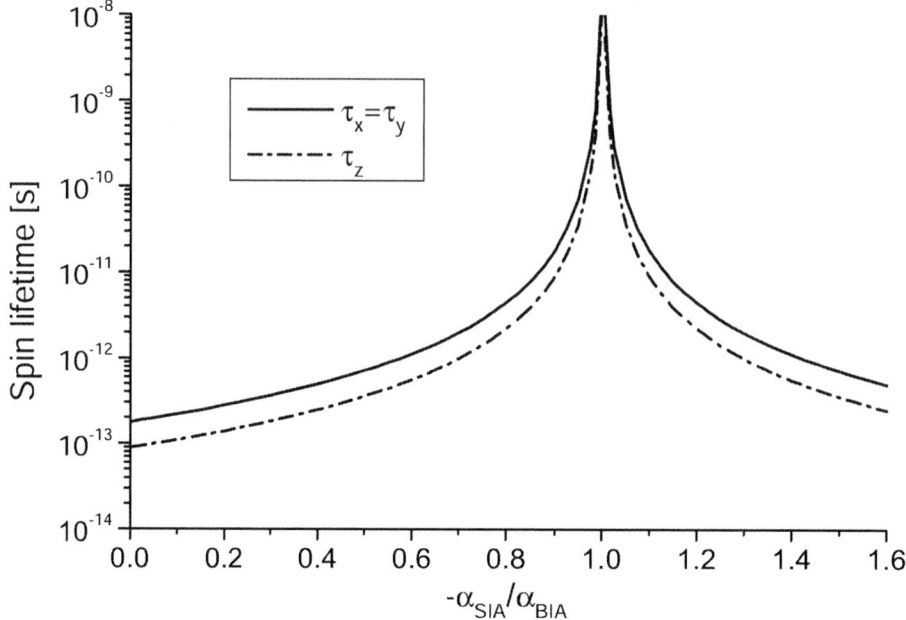

Figure 1: Spin lifetimes for the three spin components for a [111]-grown QW (upper panel) and a [110]-grown QW (lower panel) as a function of the ratio of the SIA and the BIA parameters.

have the magnetization in-plane as normally obtainable from the demagnetizing field minimization requirements.

Another advantage that must be pointed out for [111] heterostructures is that, because of the form of Eq. (2), BIA effects give only a constant background to the Rashba Hamiltonian, and therefore will not disturb, to first order, the operation of the Datta-Das transistor.

At first, an ensemble of spins is injected in the 2DEG. The gate bias drives the lifetime of the injected spins on- or off-resonance by setting $\alpha_{BIA} = -\alpha_{SIA}$ or $\alpha_{BIA} \neq -\alpha_{SIA}$, respectively. In the "on" state the spins would arrive aligned with the ferromagnetic collector, thus resulting in low resistance. In the "off" state, the spins are randomized before reaching the collector and a high resistance is measured.

There are other kinds of devices that can be constructed with these building blocks. If the gate bias in Fig. 2 is applied through a charged/uncharged floating gate, the device would behave as a flash memory. A different nonvolatile memory configuration can be obtained from Fig. 2 if the gate bias is always kept at the resonance condition. Then, the "0" or "1" states would be given by the relative orientation of the magnetization of the emitter and the collector. The performance would improve because this memory can operate in the ballistic mode. We can also envision a magnetic information readout head based on this last nonvolatile memory. Similar to giant magnetoresistance readout heads [29], the magnetization of one contact would be pinned while the other follows some stored pattern. Finally, strain effects are likely to also distort the spin lifetime of the electrons, which might lead to spintronic strain gauges.

5. SUMMARY

In summary, we have shown that electrons in the conduction band of a [111] zincblende quan-

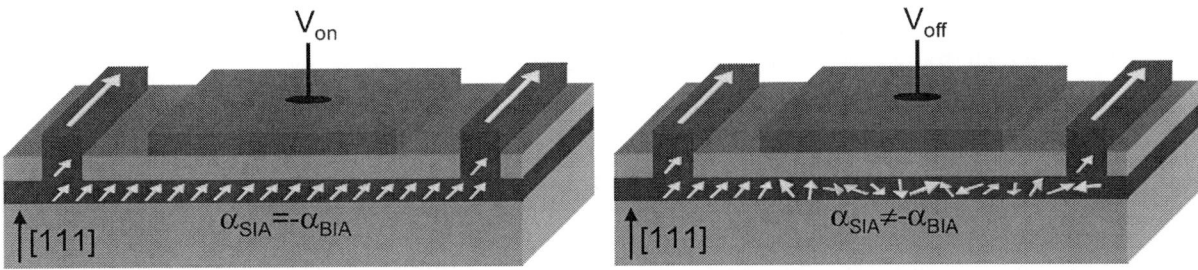

Figure 2: Operating principle of the resonant spin lifetime transistor. The gate bias drives the lifetime of the injected spins on- or off-resonance. In the "on" state the spins would arrive aligned with the ferromagnetic collector, thus resulting in low resistance. In the "off" state, the spins are randomized before reaching the collector and a high resistance is measured.

tum well have extended D'yakonov-Perel' spin lifetimes for *all* spin components when BIA and SIA effects are of equal magnitude. This effect can be used to improve the resonant spin lifetime transistor, where a gate bias modulates the resistance of a channel through the spin lifetime of a 2DEG. The [111] version of the device is free from constraints in the transport direction crystallographic orientation and it does not need to specify the orientation of the magnetization of the contacts. Also, the characteristics of the two-band spin Hamiltonian suggest that implementation of the Datta-Das spin transistor in a [111] quantum well will be much less sensitive to bulk inversion asymmetry effects.

ACKNOWLEDGMENTS

The authors thank Dr. Stuart Wolf for helpful discussions. This work was supported in part by Defense Advanced Research Projects Agency (DARPA) under Contracts No. MDA972-01-C-0002 and No. DAAD19-01-1-0324. A part of this work was carried out at the Jet Propulsion Laboratory, California Institute of Technology, through an agreement with the National Aeronautics and Space Administration.

    * Currently at: Computational Research Division, Lawrence Berkeley National Laboratory, Berkeley, CA 94720, USA

References

[1] L. Esaki, R. Tsu, *IBM J. Res. Develop.* **14**, 61 (1970).

[2] K. K. Likharev, *Proc. IEEE* **87**, 606 (1999). And references therein.

[3] S. Datta, B. Das, *Appl. Phys. Lett.* **56**, 665 (1990).

[4] Y. A. Bychkov, E. I. Rashba, *J. Phys. C* **17**, 6039 (1984).

[5] G. Dresselhaus, *Phys. Rev.* **100**, 580 (1955).

[6] M. I. D'yakonov, V. I. Perel', *Fiz. Tverd. Tel.* **13**, 3581 (1971).

[7] M. I. D'yakonov, V. Y. Kachorovskiǐ, *Fiz. Tekh. Poluprovodn.* **20**, 178 (1986). [Sov. Phys. Semicond. **20**, 110 (1986)].

[8] J. Luo, H. Munekata, F. F. Fang, P. J. Stiles, *Phys. Rev. B* **38**, 10142 (1988).

[9] G. E. Pikus, A. N. Titkov, *Optical Orientation*, F. Meier, B. P. Zakharchenya, eds. (North Holland, Amsterdam, Netherlands, 1984), vol. 8, pp. 73–131.

[10] N. S. Averkiev, L. E. Golub, M. Willander, *J. Phys.: Condens. Matter* 14, R271 (2002).

[11] J. Schliemann, J. C. Egues, D. Loss, *Phys. Rev. Lett.* 90, 146801 (2003).

[12] X. Cartoixà, D. Z.-Y. Ting, Y.-C. Chang, *Appl. Phys. Lett.* 83, 1462 (2003).

[13] N. S. Averkiev, L. E. Golub, *Phys. Rev. B* 60, 15582 (1999).

[14] A. A. Kiselev, K. W. Kim, *Phys. Stat. Sol. (b)* 221, 491 (2000).

[15] K. C. Hall, W. H. Lau, K. Gündogdu, M. E. Flatté, T. F. Boggess, *Appl. Phys. Lett.* 83, 2937 (2003).

[16] R. Eppenga, M. F. H. Schuurmans, *Phys. Rev. B* 37, 10923 (1988).

[17] X. Cartoixà, D. Z.-Y. Ting, Y.-C. Chang, *cond-mat/0402237* (2004).

[18] W. H. Lau, J. T. Olesberg, M. E. Flatté, *Phys. Rev. B* 64, 161301 (2001).

[19] T. Schäpers, G. Engels, J. Lange, T. Klocke, M. Hollfelder, H. Lüth, *J. Appl. Phys.* 83, 4324 (1998).

[20] X. Cartoixà, D. Z.-Y. Ting, E. S. Daniel, T. C. McGill, *Superlatt. Microstruct.* 30, 309 (2001).

[21] J. Nitta, T. Akazaki, H. Takayanagi, T. Enoki, *Phys. Rev. Lett.* 78, 1335 (1997).

[22] M. Cardona, N. E. Christensen, G. Fasol, *Phys. Rev. B* 38, 1806 (1988).

[23] X. Cartoixà, D. Z.-Y. Ting, T. C. McGill, *cond-mat/0212394* (2002).

[24] R. J. Elliott, *Phys. Rev.* 96, 266 (1954).

[25] Y. Yafet, *Solid State Physics*, F. Seitz, D. Turnbull, eds. (Academic Press, New York, 1963), vol. 14, pp. 1–98.

[26] G. L. Bir, A. G. Aronov, G. E. Pikus, *Zh. Eksp. Teor. Fiz.* 69, 1382 (1975). [Sov. Phys. JETP 42, 705 (1976)].

[27] Y. Ohno, R. Terauchi, T. Adachi, F. Matsukura, H. Ohno, *Phys. Rev. Lett.* 83, 4196 (1999).

[28] O. Z. Karimov, G. H. John, R. T. Harley, W. H. Lau, M. E. Flatte, M. Henini, R. Airey, *Phys. Rev. Lett.* 91, 246601 (2003).

[29] C. Tsang, R. E. Fontana, T. Lin, D. E. Heim, V. S. Speriosu, B. A. Gurney, M. L. Mason, *IEEE Trans. Magn.* 30, 3801 (1994).

Cambridge University Press
32 Avenue of the Americas, New York, NY 10013-2473, USA

Materials Research Society
506 Keystone Drive, Warrendale, PA 15086

ISBN 978-1-55899-753-0